T0258035

Advances in
Ceramic Armor VI

Advances in Ceramic Armor VI

A Collection of Papers Presented at the
34th International Conference on Advanced
Ceramics and Composites
January 24–29, 2010
Daytona Beach, Florida

Edited by
Jeffrey J. Swab

Volume Editors
Sanjay Mathur
Tatsuki Ohji

A John Wiley & Sons, Inc., Publication

Published by John Wiley & Sons, Inc., Hoboken, New Jersey.
Published simultaneously in Canada.

For general information on our other products and services or for technical support, please contact our
Customer Care Department within the United States at (800) 762-2974, outside the United States at
(317) 572-3993 or fax (317) 572-4002.

Wiley also publishes its books in a variety of electronic formats. Some content that appears in print may
not be available in electronic format. For information about Wiley products, visit our web site at
www.wiley.com.

Library of Congress Cataloging-in-Publication Data is available.

ISBN 978-0-470-59470-4

Printed in the United States of America.

10 9 8 7 6 5 4 3 2 1

Contents

Preface

The Armor Ceramics Symposium was held January 25–27, 2010 in Daytona Beach, FL as part of the 34th International Conference and Exposition on Advanced Ceramics and Composites. The 8th edition of this symposium consisted of over 65 oral and poster presentations on topics such as Impact, Penetration and Material Modeling, Boron Carbide, Silicon Carbide, Dynamic Material Behavior, Transparent Materials, and NDE Applications. The symposium continues to foster discussion and collaboration between academic, government and industry personnel from around the globe.

On behalf of the organizing committee I would like to thank the presenters, authors, session chairs and manuscript reviewers for their efforts in making the symposium and the associated proceedings a success. As special thanks goes to Marilyn Stoltz and Greg Geiger of The American Ceramic Society who was always there to answer my questions and provide the guidance and administrative support necessary to ensure a successful symposium.

JEFFREY J. SWAB

Introduction

This CESP issue represents papers that were submitted and approved for the proceedings of the 34th International Conference on Advanced Ceramics and Composites (ICACC), held January 24–29, 2010 in Daytona Beach, Florida. ICACC is the most prominent international meeting in the area of advanced structural, functional, and nanoscopic ceramics, composites, and other emerging ceramic materials and technologies. This prestigious conference has been organized by The American Ceramic Society's (ACerS) Engineering Ceramics Division (ECD) since 1977.

The conference was organized into the following symposia and focused sessions:

Symposium 1	Mechanical Behavior and Performance of Ceramics and Composites
Symposium 2	Advanced Ceramic Coatings for Structural, Environmental, and Functional Applications
Symposium 3	7th International Symposium on Solid Oxide Fuel Cells (SOFC): Materials, Science, and Technology
Symposium 4	Armor Ceramics
Symposium 5	Next Generation Bioceramics
Symposium 6	International Symposium on Ceramics for Electric Energy Generation, Storage, and Distribution
Symposium 7	4th International Symposium on Nanostructured Materials and Nanocomposites: Development and Applications
Symposium 8	4th International Symposium on Advanced Processing and Manufacturing Technologies (APMT) for Structural and Multifunctional Materials and Systems
Symposium 9	Porous Ceramics: Novel Developments and Applications
Symposium 10	Thermal Management Materials and Technologies
Symposium 11	Advanced Sensor Technology, Developments and Applications

Focused Session 1 Geopolymers and other Inorganic Polymers
Focused Session 2 Global Mineral Resources for Strategic and Emerging
                   Technologies
Focused Session 3 Computational Design, Modeling, Simulation and
                   Characterization of Ceramics and Composites
Focused Session 4 Nanolaminated Ternary Carbides and Nitrides (MAX Phases)

The conference proceedings are published into 9 issues of the 2010 Ceramic Engineering and Science Proceedings (CESP); Volume 31, Issues 2–10, 2010 as outlined below:

- Mechanical Properties and Performance of Engineering Ceramics and Composites V, CESP Volume 31, Issue 2 (includes papers from Symposium 1)
- Advanced Ceramic Coatings and Interfaces V, Volume 31, Issue 3 (includes papers from Symposium 2)
- Advances in Solid Oxide Fuel Cells VI, CESP Volume 31, Issue 4 (includes papers from Symposium 3)
- Advances in Ceramic Armor VI, CESP Volume 31, Issue 5 (includes papers from Symposium 4)
- Advances in Bioceramics and Porous Ceramics III, CESP Volume 31, Issue 6 (includes papers from Symposia 5 and 9)
- Nanostructured Materials and Nanotechnology IV, CESP Volume 31, Issue 7 (includes papers from Symposium 7)
- Advanced Processing and Manufacturing Technologies for Structural and Multifunctional Materials IV, CESP Volume 31, Issue 8 (includes papers from Symposium 8)
- Advanced Materials for Sustainable Developments, CESP Volume 31, Issue 9 (includes papers from Symposia 6, 10, and 11)
- Strategic Materials and Computational Design, CESP Volume 31, Issue 10 (includes papers from Focused Sessions 1, 3 and 4)

The organization of the Daytona Beach meeting and the publication of these proceedings were possible thanks to the professional staff of ACerS and the tireless dedication of many ECD members. We would especially like to express our sincere thanks to the symposia organizers, session chairs, presenters and conference attendees, for their efforts and enthusiastic participation in the vibrant and cutting-edge conference.

ACerS and the ECD invite you to attend the 35th International Conference on Advanced Ceramics and Composites (http://www.ceramics.org/icacc-11) January 23–28, 2011 in Daytona Beach, Florida.

Sanjay Mathur and Tatsuki Ohji, Volume Editors
July 2010

INSPECTING COMPOSITE CERAMIC ARMOR USING ADVANCED SIGNAL PROCESSING TOGETHER WITH PHASED ARRAY ULTRASOUND

J. S. Steckenrider
Illinois College
Jacksonville, IL

W. A. Ellingson and E.R. Koehl
Argonne National Laboratory
Argonne, Illinois USA

T.J. Meitzler
US Army, TARDEC
Warren, MI

ABSTRACT

A series of 16-inch square by 2-inch thick, multi-layered ceramic composite armor specimens have been inspected using a 128 element, 10MHz immersion phased array ultrasound system. Some of these specimens had intentional design defects inserted interior to the specimens. Because of the very large changes in acoustic velocities of the various layered materials, ultrasonic wave propagation is problematic. Further, since the materials used in the layers were stacked such that a lower elastic modulus material was on one side and a higher elastic modulus material was on the other, the side selected for ultrasonic insonification became a significant parameter. To overcome some aspects of the issues with the acoustic wave propagation, two digital signal processing methods were employed. These were: 1)- use of fast Fourier transforms (FFT) and 2)-an integrated signal analysis. Each method has strengths and weaknesses with application in part dependent upon the side of sample used for insonification. The results clearly show that use of these methods significantly improves defect detection. This paper presents the details of the samples used, the issues with ultrasonic wave propagation, a discussion of the two digital signal processing algorithms and results obtained.

INTRODUCTION

Ceramic armor for vehicles offers significant potential improvement over historical materials as it provides a greater capacity for energy absorption and dissipation per unit mass, achieved through very high fracture toughness. However, unlike their metallic predecessors, ceramic materials are much more vulnerable to manufacturing defects, as any such flaws can dramatically reduce that high toughness, thereby compromising the armor's ability to protect military personnel. Thus, an efficient non-destructive evaluation (NDE) method which can identify these defects before the armor is placed into service is critical to their effectiveness[1]. Conventional ultrasonic techniques have been used to both locate and characterize such defects in the monolithic ceramic tiles that make up the "backbone" of these armor panels[2,3]. Furthermore, phased-array ultrasound[4] (PA-UT) has demonstrated significant improvement over these methods[5] as it offers both enhanced sensitivity and markedly improved throughput[6,7].

While PA-UT has clearly demonstrated its performance with regard to the monolithic tiles themselves, actual implementation of ceramic armor incorporates these monolithic tiles into a thick, multi-layered ceramic composite structure. This creates an additional manufacturing step during which additional defects may be introduced, and thus an additional NDE screening would be required to reduce ballistic performance failures caused by these new defects. However, because of the composite nature of the armor panels, the same NDE methodologies used for the monolithic tiles cannot simply be re-applied, as the composite design brings with it additional challenges that must be addressed in order to validate the integrity of the panels.

APPROACH

Figure 1a shows a schematic diagram of the layered structure of the composite armor panels. As shown, there are four primary constituent materials used in the panel makeup: the high-toughness ceramic tiles, a carbon-based matrix used to encase these tiles, an elastomeric layer used to distribute and attenuate mechanical stresses transmitted by the ballistic impact, and a glassy layer which provides a monolithic substrate to support the composite armor. The individual tiles are then arranged side-by-side to form a panel, as shown in Figure 1b where a regular hexagonal tessellation (as was used in this work) is shown.

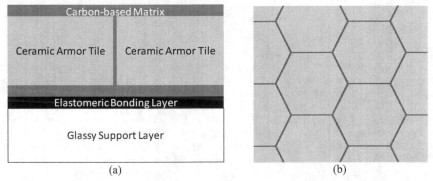

(a)   (b)

Figure 1.   Schematic diagram of composite layered armor panel: a) cross-section and b) top view.

For the test panels investigated in this effort, two modifications were made to the as-designed panel layup. First, planar inclusions were intentionally inserted at the two most critical boundaries (i.e., on either side of the elastomeric bonding layer) to simulate a "disbond" at the locations where it would have the greatest effect in reducing the ballistic performance of the panel. Second, two different armor tiles were used, each made from a different material, so as to reduce the manufacturing investment in each panel. The more resource-intensive material (which also has the greater fracture toughness) was used in the tiles immediately adjacent to the inclusion, while all other tiles were made from the alternate material. Two different configurations for the included "disbond" were used: one in the center of a tile and one at the intersection of three tiles, as shown in Figure 2. Finally, three different "disbond" sizes were used (0.5", 1.5" and 2.5" diameters).

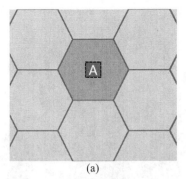
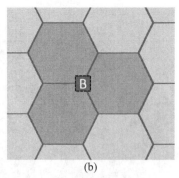

(a)	(b)

Figure 2. Schematic diagram of composite layered armor panel showing pattern of ceramic tiles (where the darker tile is made from the more resource-intensive material) and location of intentional defects for a) single-tile location and b) tile intersection location.

To evaluate the integrity of these elastomer boundaries the panels were inspected using a water-immersion phased-array system. The phased-array probe consisted of 128 transducer elements operating at 10 MHz with an active area of 64mm by 7 mm. By using a phased-array system we were able to dramatically improve inspection time (by eliminating most of the motion in the scan axis, replacing it with the much faster electronic indexing of the phased array, as illustrated in Figure 4) and adjust probe focus to accommodate the more complex structure of these panels. Data acquired with this method yields a 3-D volume of reflection data. The data are collected such that a single amplitude value is recorded for each point in the inspected 3-D volume. For the 16" x 16" panels used in this study, such single point data collection amounts to over 180 Mb of data per panel when data acquired with 1 mm lateral resolution at a 50 MHz data acquisition rate. (Note that all c-scan images presented herein show a 16" x 13.25" region of the panel.) However, because the acoustic impedance of the elastomeric material is dramatically different than that of the carbon based matrix, very little of the incident acoustic energy is able to penetrate through this layer. Therefore, each panel was inspected twice, once from each side. Given the differences in properties and layup on either side of the elastomer, different inspection methodologies were used on either side.

In the general case for immersion scanning ultrasonic methods (be it phased array or single transducer), when the exact location/depth of a defect(s) is not known, a 2-D representation of the scanned data can be obtained in the form of a c-scan, or 2-D map of peak amplitude as a function of scan and index position across the face of the plate. Usually this C-scan information is produced by selecting the peak amplitude from the reflections within the material, independent of depth. To provide some depth information for that peak amplitude pulse, a time-of-flight (TOF) c-scan often accompanies an amplitude c-scan. The arrival time of that peak, for a monolithic specimen, is linearly correlated to the depth from which that peak amplitude was reflected. However, if the plane or depth of the defect is well known then only the reflected pulse data from that plane are of interest. In such a case, in the A-scan data, a "time-width gate" can be set and ONLY data from within that time window will be used for the resulting C-scan. The size of the digital data set in such a case becomes much more manageable. Examples of conventional C-scan results for one of the armor panels are shown in Figure 3 below, in which timing gates were set around the first, second and third reflections from the Matrix/Elastomer interface.

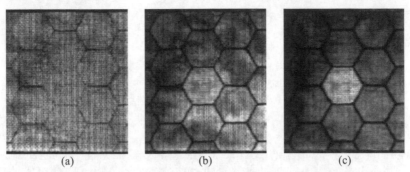

(a)   (b)   (c)

Figure 3. Conventional C-scan images of reflection from the Matrix/Elastomer boundary with timing gates centered on a) first, b) second and c) third reflections.

When the specimen to be studied by ultrasound is not monolithic, e.g., is a layered medium such as these composite armor samples, or if a sample contains a number of distinct closely "stacked" defects (i.e., multiple defects that occur along a line normal to the inspection surface) then the ultrasound process becomes more complicated. In such a case, the issue becomes a matter of trying to distinctly identify one layer separate from the others. In this case, the conventional C-scan approach, which is only able to capture information for a single location/depth for each point on the inspected surface, is often unable to specifically identify/locate a defect unless it happens to coincide with the most reflective interface, and is generally unable to distinguish between multiple collinear defects. Furthermore, when only a single plane is of interest, and the defect(s) of interest coincides with that plane, reflections from other interfaces in a layered medium may overlap that response, masking the effect of the defect. In these cases, the defect signature is often still present within the full data set, but extracting that signature using automated methods becomes a challenge. Insonification from the top surface of an armor panel generates multiple reflections, as demonstrated in Figure 4a. While the timing control afforded by the phased-array system naturally reduces the effect of interfaces near the surface, those close to the interface of interest are relatively in-phase with reflections from this interface, and thus can contribute significant noise. This is seen in Figure 4b in which the discrete reflections from the Matrix/Elastomer interface (the 6 narrow dark bands labeled R1 through R6) are interspersed with additional reflections from other interfaces (as indicated by the additional gray bands throughout the B-scan, especially in the top half). With regard to the issue of multiple layer reflections, the current effort has evaluated two filtering methodologies which significantly improve the signal-to-noise ratio over conventional methodologies. These filtering techniques have been applied to the received signals from ceramic-side inspections (where the multiple layers of high-velocity materials give rise to many interfering reflections), but are not of particular use for signals from support layer side inspections (where the tile is essentially monolithic down to the elastomer-matrix interface, and very little acoustic energy is reflected from interfaces beyond that one).

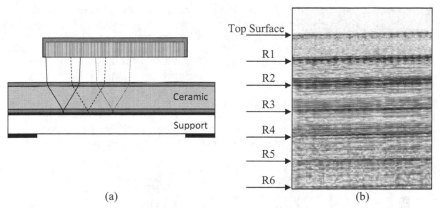

(a)   (b)

Figure 4.   a) Schematic diagram of phased-array inspection of layered armor panel and b) resulting B-scan plotting reflected acoustic amplitude as a function of depth (vertical axis) and position on the panel (horizontal axis).

RESULTS

Graphite-side Inspection

FFT filtering: Each interface (Water/Matrix, Matrix/Ceramic, Ceramic/Matrix, Matrix/Elastomer, Elastomer/Support and Support/Water) will reflect a portion of the acoustic energy incident thereon. The relative amplitude of that reflection is dependent upon the acoustic impedance mismatch between the two materials forming that interface. Thus, while there will be reflections from all five interfaces, the reflections from the first (Water/Matrix, which reflects ~80% of the incident energy in this case) and fourth (Matrix/Elastomer, which reflects ~75% of the incident energy in this case) interfaces will have the largest amplitudes. Unfortunately, the propagation time between the first, Water/Matrix, and second, Matrix/Ceramic, reflections (as well as between the third Ceramic/Matrix and fourth, Matrix/Elastomer) will be fairly small (a fraction of a microsecond), given the relatively high velocity in the relatively thin matrix layers. Because the time duration of each reflected acoustic pulse is of the same order as the separation between the reflections from these interfaces, the reflections overlap in time. This is true of the first and second reflections from these interfaces (where the acoustic pulse is reflected between the subsurface interface, the surface of the tile, and the subsurface interface a second time). Not until the third and subsequent reflections from each of these interfaces are they sufficiently temporally separate as to be able to distinguish them (see the A-scan of Figure 5 below). However, the acoustic velocity is independent of amplitude and thus the temporal spacing between higher-order reflections is constant. Therefore, if we select only those components of the reflected acoustic signature that are periodic at this temporal spacing we should be able to greatly enhance the signal-to-noise ratio (SNR) for that interface. By examining the acoustic waveform in the frequency domain we can simply select only that frequency that corresponds to this temporal period (i.e., at the resonance frequency of the Matrix/Ceramic/Matrix "sandwich"), and the amplitude of that frequency component will be directly related to the reflectance of the interface of interest. Using this amplitude allows plotting scans for all positions in the "FFT Filtered" -scans. An example for one such scan for panel D3 is shown in Figure 7a below. Interestingly, one additional benefit of this approach is the segregation of tile types. Because the two grades of tile used in these panels have different acoustic velocities, they also have different resonance frequencies. By selecting the frequency corresponding to the higher-performing tile material (with the higher acoustic velocity), only that tile type contributes significantly to the FFT-filtered C-scan of Figure 7. This is further illustrated in the B-scan of Figure 6a below which shows the different time/depth spacing of the two tile types (with the higher-performance tile in the center). One can then alternatively select the frequency corresponding to the alternate tile type to arrive at the FFT-filtered C-scan of Figure 6b, which not only emphasizes the alternate tile regions, but also reveals a C-shaped defect in the tile just above the center tile.

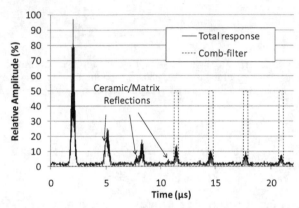

Figure 5. Schematic diagram of a theoretical A-scan showing reflections from interfaces within a multiple layered-structure. For clarity, the waveform shown includes only the components of the waveform due to the Ceramic/Matrix and the Matrix/Elastomer interfaces.

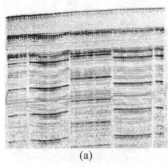 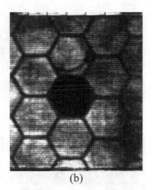

(a)   (b)

Figure 6.   a) B-scan of Ceramic-side inspection showing the different resonances for different
tile materials, b) FFT-filtered C-scan tuned to examine the lower-performance tile
regions.

Comb filtering: While the FFT filtering has the advantages of being very straightforward
(requiring only velocity and thickness information for each material used) and including all
transits between the Water/Matrix and Matrix/Elastomer interfaces (including those initially
generated by other reflections), it still includes the initial front-surface reflection and the "ring-
down" from the first Ceramic/Matrix interface reflection in its analysis, and thus may not
maximize the SNR (as these reflections do not contain any data about the interface in question).
This is reflected in the slightly stronger weave pattern shown in the FFT results of Figure 7a (as
compared with the Comb filter results of Figure 7b). A more direct approach was therefore
attempted. This approach required one additional piece of information – the arrival time of the
Water/Matrix interface front-surface reflection. Once that is known, the exact arrival times for
all subsequent reflections can be determined (again – if the velocity and thickness information
for each layer is known) in a "front-follow" arrangement. A "comb-filter" can then be
constructed to pass only those components of the acoustic waveform that correspond to the
arrival times of reflections from the interface of interest, beginning with the third (where
temporal separation from other reflections begins). By integrating the response within each of
these reflections, the comb-filtered waveform is able to distinguish the reflection from the
interface of interest, thereby maximizing the phased—array reflection pulse sensitivity to defects
at that interface, as shown in Figure 7b. However, it is worth noting that, while neither of these
images reveal the included artificial "disbond" (as might be expected, given the already very
high mismatch in impedance between the Matrix and Elastomer layers), there is clearly non-
uniformity in the amount of acoustic energy reflected from this interface. The high-reflectivity
regions on the interface are more clearly evident using the comb-filter technique, while the low-
reflectivity regions are better visualized using the FFT method. Since we do not yet know which
is a better predictor of ballistic performance, data was acquired for all samples using both
techniques.

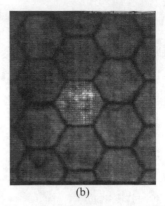

(a)   (b)

Figure 7.   Phased array ultrasound data filtered c-scan data for tile D3, inspected from the
graphite-side, using the a) FFT analytical method and b) comb-filter technique.

S2-side Inspection
        Insonification of the armor tile from the Support side of the tile presents essentially a
monolithic material, as the acoustic impedance of the Support and Elastomer materials are
virtually identical. A conventional c-scan approach, properly employed, can therefore be applied
for inspection of the Support/Elastomer interface, as this interface will typically be the most
reflective plane within the layered structure (as much of the acoustic energy incident on the
Matrix side of the Elastomer will be totally internally reflected within the Elastomer). However,
if there are any smaller defects present within the thickness of the Support layer a conventional
c-scan approach will be insensitive to these defects. Thus, for our results from these inspections,
two c-scan images are provided. The first is a conventional c-scan in which the phased array is
focused on the Elastomer/Matrix interface for both transmission and reception. The second
specifically excludes this interface from the resulting a-scans and examines the remainder of the
volume of the Support and Elastomer materials for either additional discontinuities within the
Support material, or degradation of the Support/Elastomer interface. In either case, any defect
revealed can be further analyzed using the full data set. An example of this approach is shown in
Figure 8 below. Figure 8a shows a B-scan (essentially a vertical slice through the 180 Mb data
set) of the Support material in panel C5. The Support/Elastomer interface is barely
distinguishable at a "depth" of approximately 13 µs (given the virtually identical impedances for
these two layers, less than 1% of the energy would be reflected from an intact interface).
However, there is also a much more pronounced intermediate reflection at an apparent "depth"
of approximately 11 µs – within the Support material. For reference, the same cross-sectional
"slice" from panel C6 is shown in Figure 8b. Here, the Support/Elastomer interface is not really
visible at all at 13 µs, but there is no intermediate reflection at 11 µs. Instead, though a reflection
is seen at approximately 5 µs. Figure 9a shows a c-scan of the Support and Elastomer material
layers in panel C5, where a reduced reflection is clearly evident (in the form of a thumbprint-
shaped region protruding from the lower left quadrant of the image), indicating that there is a
higher reflection (i.e., a defect) covering the entire sample outside the thumbprint-shaped region.
The TOF scan in Figure 9b shows that the peak amplitude within this thumbprint-shaped region
is coming from near the Elastomer/Matrix interface, but that the peak signal in areas outside this

region comes from within the Support material itself. This would therefore indicate that a "disbond" within the Support layer has occurred everywhere outside the thumbprint region. Thus, if detection and location of these non-interfacial defects are of interest, the inspection methodology must be able to maximize sensitivity to all possible depths within the Support material, not just the Support/Elastomer interface.

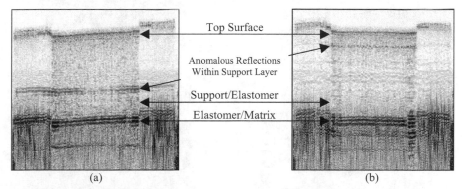

(a)   (b)

Figure 8. 10 MHz Phased array B-scan data for S2-side insonification of tiles a) C5 and b) C6.

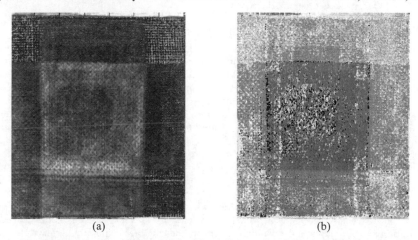

(a)   (b)

Figure 9. Phased array C-scan data for tile C5. a) Conventional C-scan of the Support and Elastomer, showing a thumbprint-shaped reduced reflection region, and b) TOF image, showing a uniform time/depth of 11 μs below the surface for the area outside this region

In cases where an artificial "disbond" was introduced between the Support and Elastomer layers, one would expect to see a significant change in the reflected energy, as this defect would marked alter the acoustic impedance mismatch of that interface. This is clearly seen in the Band C-scans of panel J3, shown in Figure 10 below. The B-scan of Figure 10a was taken along a vertical line through the center of the panel, bisecting the ½" square defect. Of note is the fact that, in addition to revealing the defect, the B-scan also indicates that the overall integrity of the Support/Elastomer interface is not as strong as was observed in the earlier panels of Figure 8.

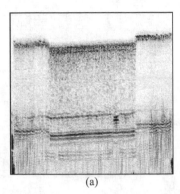
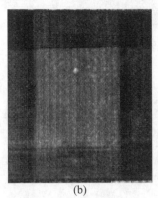

(a) (b)

Figure 10. a) B-scan and b) C-scan of panel J3 clearly indicating the included defect within the Support/Elastomer interface.

CONCLUSION

In summary, two specific digital signal processing techniques have been developed to help improve phased array ultrasonic inspection and analysis of multi-layered ceramic armor panels. The location of some of the specific layered materials, especially the highly attenuative central Elastomer layer, suggests that ultrasonic inspections for locating and identifying internal defects be conducted from each side of the panels. In addition, because of the specific properties of the layers, different inspection methodologies should be used for each side in order to maximize detection sensitivity. Inspection from the Ceramic side, where the multiple layers of high-velocity materials create multiple and often overlapping acoustic reflections, use of an FFT or Comb-filtering technique likely would be best to locate and identify interfacial defects. The detection sensitivity from these two methods differs slightly. Comparing these two techniques to determine which of the two is the "better" method depends upon the predictability of ballistic performance provided by each technique.

Unlike the Ceramic-side insonification, where multiple high-velocity layers are present, inspection from the Support side, in which the panel behaves largely as a monolithic material, can be accomplished without the additional signal processing. Detection of interfacial defects at the Support/Elastomer interface has been clearly demonstrated. However, in cases where defects occur within the Support layer (a particularly likely scenario for post-impact examination, but still observed even within the untested panels), it is critical not only that these defects be detected (even when multiple "stacked" defects are present) but that the sensitivity of the detecting system be maximized for that detection. This would best be performed using the Dynamic Depth Focusing (DDF) method unique to phased-array inspection. In this method, the acoustic energy is transmitted into the material with a time delay between array elements that will "focus" the energy at a particular depth (say, the Support/Elastomer interface). If only interfacial defects are of interest, a simple c-scan of the reflection from this interface is sufficient to identify and locate those defects. However, if defects within the bulk of the Support material are also of interest (as is clearly the case for the panels presented here, in which there appears to be some kind of planar defect within the Support layer), DDF allows for the time delay between array elements to be adjusted dynamically so that the *received* signals are "focused" at each reflecting depth rather than the transmitted depth only (as would be the limitation for conventional ultrasound). This greatly enhances both the sensitivity to and fidelity of non-interfacial defects without compromising sensitivity to interfacial defects, and thus will be the focus of our continuing efforts, particularly as we examine the panels after ballistic impact.

REFERENCES

[1] J.M. Wells, W.H. Green and N. L. Rupert—""On the Visualization of Impact Damage in Armor Ceramics", *Eng. Sci. and Eng. Proc.*, **22** (3), H.T. Lin and M. Singh, eds, pp. 221-230 (2002).

[2] J. S. Steckenrider, W. A. Ellingson, and J. M. Wheeler "Ultrasonic Techniques for Evaluation of SiC Armor Tile," in *Ceram. Eng. and Sci. Proc.*, **26** (7), pp. 215-222 (2005).

[3] R. Brennan, R. Haber, D. Niesz, and J. McCauley "Non-Destructive Evaluation (NDE) of Ceramic Armor Testing," in *Ceram. Eng. and Sci. Proc.*, **26** (7), pp. 231-238 (2005).

[4] G.P. Singh and J. W. Davies, "Multiple Transducer Ultrasonic Techniques: Phased Arrays" In Nondestructive Testing Handbook, 2nd Ed., **7**, pp. 284-297 (1991).

[5] J. Scott Steckenrider, William A. Ellingson, Rachel Lipanovich, Jeffrey Wheeler, Chris Deemer, "Evaulation of SiC Armor Tile Using Ultrasonic Techniques," *Ceram. Eng. and Sci. Proc.*, **27** (7), (2006).

[6] D. Lines, J. Skramstad, and R. Smith, " Rapid, Low-Cost, Full-Wave Form Mapping and Analysis with Ultrasonic Arrays", in Proc. 16th World Conference on Nondestructive Testing, September , 2004.

[7] J. Poguet and P. Ciorau, " Reproducibility and Reliability of NDT Phased Array Probes", in Proc. 16th World Conference on Nondestructive Testing, September , 2004.

A COMPARISON OF NDE METHODS FOR INSPECTION OF COMPOSITE CERAMIC ARMOR

W. A. Ellingson and E.R. Koehl
Argonne National Laboratory
Argonne, Illinois USA

T.J. Meitzler and L.P. Franks
US Army, TARDEC
Warren, MI

J. S. Steckenrider
Illinois College
Jacksonville, IL

ABSTRACT

A US Army-supported project, Effects of Defects, has been established to help identify appropriate NDE modalities for inspecting layered ceramic-composite armor. As a part of this effort, a series of 84 40 cm (16-inch) square by 50 mm (2-inch) thick, multi-layered ceramic-composite armor specimens has been prepared and will be "inspected" using an array of Nondestructive Evaulation (NDE) methods. Some of the test panels have had "designed defects" located in the interior—some do not. All samples are to be ballistically impacted and are to be inspected before and after ballistic testing. The question to be answered is—which NDE modality might best be used to quantify ballistically-induced damage. NDE modalities under present study include: 1)-immersion phased array ultrasonics, 2)- through-transmission, direct-digital x-ray imaging, 3)-non-contact scanning microwaves, 4)-air-coupled ultrasound and 5)-immersion, through-transmission and pulse echo single-transducer ultrasound. At this time, all 84 samples have been inspected prior to ballistic testing. This paper will discuss details about the project, present an overview of the NDE techniques, discuss issues that have been uncovered and briefly present results that have been obtained.

INTRODUCTION

Effectiveness of multi-layer composite-ceramic armor appliqués, that are to be mounted on military vehicles, can be degraded by defects within the ceramic from production and/or by damage resulting from handling or impact from ballistic threat projectiles. Detection of defects during production is necessary to assure reliable and cost-effective armor production. Detection of usage-induced damage is necessary to determine the integrity of in-theatre armor so that appropriate and timely replacement can be made. Recently, an "effects of defects" project was undertaken by the US Army to assess several NDE technologies for detection of damage. The NDE methods under evaluation include: 1)-immersion phased array ultrasonics, 2)- through-transmission, direct-digital x-ray imaging, 3)-non-contact scanning microwaves, 4)-air-coupled ultrasound and 5)-immersion, through-transmission and pulse-echo single-transducer ultrasound. In order to evaluate these NDE methods, a set of 84 specially made ceramic-composite samples was produced. Shown below in Figure 1 is a schematic diagram of a cross-section of a multi-layered ceramic-composite armor used in this effort. To be noted is that ceramic tile is "sandwiched" between two thinner layers of special material and this "sandwich' is followed by a layer of an elastomer and then followed by a thicker layer of a low-elastic modulus material. From the NDE standpoint, such an armor design presents challenges for inspection. The ballistic impact side is the "sandwich' side and the low elastic modulus material side is the side that is mounted to the vehicle. Table I shows the list of samples along with schematic diagrams of the various designs.

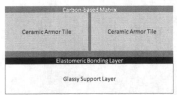

Figure1. Schematic diagram of cross section of multi-layer
ceramic composite armor

Table I   List and schematic diagrams of NDE Effects of Defects samples

Run Designation	Panel Designation	Panel Qty	Flaw Size	Flaw Position	Center/Triple Pt
A	A1-A4	8	no flaws		
B	B1-B4	8	no flaws		
C	C1-C4	8	½" square		
D	D1-D4	8	1½" square		
E	E1-E4	8	2½" square		
F	F1-F4	8	½" square		
G	G1-G4	8	1½" square		

H	H1-H4	8	2½" square		
J	J1-J4	8	½" square		
K	K1-K4	8	1½" square		
L	L1-L4	8	2½" square		
M	M1-M4	8	½" square		
N	N1-N4	8	1½" square		
O	O1-O4	8	2½" square		

To be noted is that the inserted "defects" are placed either above the layer immediately behind the ceramic insert or placed below the elastomer layer following the ceramic insert. In addition, the sizes of the defects inserted varied from 12 mm square (1/2-inch) to 62 mm square (2 ½-inches). Fabrication of the armor test panels further utilized two different ceramic materials. A slightly higher density and different composition material was used in the immediate vicinity of the inserted defects, (see gray region in schematics of Table I), and a lower density ceramic material was then used for all surrounding tile (the yellow regions in the schematics of Table I). In this project, the NDE methods to be explored were not restricted to those with the potential for "on-vehicle' inspection. Rather the NDE methods selected were those that seemed appropriate for inspection of as-produced armor appliqués and appliqués that have been removed from vehicles. It is the overall plan to develop NDE capability to effectively estimate the flaw size and location in ceramic armor in order to support survivable structures for Ground Combat Vehicles (GCV's).

PLAN/SCOPE OF WORK

The scope of work for the entire effort actually has nine tasks and the effort being reported on here is only one aspect of this entire project. The nine tasks in the project are defined as follows: 1) Manufacture multi-hit test panels, 2) Conduct initial NDE studies, 3) Manufacture 50 multi-hit test panels with and without intentional defect4) Conduct initial NDE prior to ballistic impact testing, 5) Ballistically impact test panels, 6) Manufacture 112 single shot panels, 7) Conduct "best-effort" NDE studies of single shot panels prior to ballistic impact tests, 8) Ballistically impact 84 of the 112 single shot test panels, and 9) Conduct radiography NDE tests on all 84 ballistically impacted test panels.

The fabrication of the intentional defect test panels within the 84 panel single-shot group is a rather complex set of ceramic processing steps that has required extensive development. An example of what a partially completed test panel looks like is shown in Figure 2. The photograph shows one of the test panels that is a "'triple point" sample and has three silicon carbide ceramic test blocks surrounded by aluminum oxide test blocks. These panels were then cut to yield four test panels each with three silicon carbide blocks in the center as noted in Table I.

Figure 2: Photograph of intentional defect test panel during fabrication. The three darker tiles shown in four locations are silicon carbide tiles used for the 'triple point" samples

DESCRIPTION OF THE NDE METHODS

The following section briefly describes the nondestructive evaluation (NDE) methods being explored in this effort. These NDE methods include: 1)-immersion phased-array ultrasonics, 2)- through-transmission, direct-digital x-ray imaging, 3)-non-contact scanning microwaves, 4)-air-coupled ultrasound and 5)-immersion, through-transmission and pulse-echo single-transducer ultrasound.

1. Immersion Phased-Array Ultrasonics

Phased-array ultrasonic methods[1] have been discussed previously for application to ceramic armor, but all previous work was only on the ceramic material itself. In a layered structure with several different materials, inherently different acoustic velocities cause refraction of the acoustic wave, and the effects on defect detection are unknown. However, previous work[1] on defect detection in armor quality ceramics had clearly demonstrated that a much higher signal to noise ratio (S/N) was obtained for a phased-array system as compared to a single-transducer immersion ultrasound system. Shown below in Figure 3 are photographs of the immersion phased-array ultrasound system used in this study. This system can drive a phased-array transducer with up to 128 individual elements with control of the sequencing of up to 32 elements at any one time. Thus this is a 32/128 system.

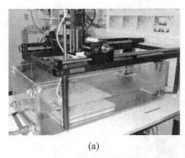
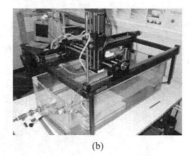

(a)   (b)

Figure 3 Photographs of immersion phased-array ultrasound equipment

The importance of using phased array technology for inspecting these ceramic composite armor panels is that by selecting the firing sequence of the transducer elements, the depth of the focus can be dynamically changed. This is shown schematically in Figure 4. By focusing at a well defined depth within the sample, the signal to noise ratio of the reflected pulse can be significantly increased.

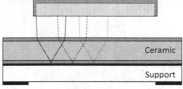

Figure 4 Schematic diagram showing firing sequence of the phased-array transducer with insonification from the "ceramic" side with the focus at the matrix to elastomer interface (horizontal dimensions not to scale).

Although inspection insonification from both ballistic impact side (as shown here) and armor mounting side have been conducted, page limitations suggest that one set of data be presented and thus all data presented herein has been obtained with insonification from from the ballistic impact side.

While the exact acoustic velocities for each material in the layered structure were unavailable, estimates were used to set the phased array transducer element firing sequence. These were determined ultrasonically by unfocused pulse-echo insonification in which the time-of-arrival of the reflection from each interface was combined with the nominal thickness of each layer to determine each layer's acoustic velocity. The materials so characterized include the carbon-based matrix (~9000 m/s), the ceramic armor material (~12,000 m/s), the elastomeric bonding layer (~1,700 m/s) and the glassy support layer (~2,300 m/s). These data were then used to set the protocol for the phased array testing. Table II presents part of the protocol.

Table II   Phased Array Protocol. Using such a protocol, the scan time for each panel was approximately 6 minutes.

•	Transducer, 128-element 10 MHz array with a 0.5 mm pitch and 7 mm width (for a total active area of 64 mm x 7 mm).
•	Conducted with 32 active elements
•	Use an 18 mm water path
•	Scan using a 1 mm (0.04") resolution in the scan direction, 0.5 mm (0.02") resolution in the index direction
•	Set scan speed to 5 mm per second (0.21 inches per second)

2. Through Transmission, Direct-Digital, X-ray Imaging

Two direct digital through-transmission x-ray imaging systems are being used in this work. In the first, the x-ray imaging work for these test panels is being conducted using a 420 kVp x-ray head coupled to a large area, 17-inch by 17-inch, flat panel detector. This set-up is shown in Figure 5. The flat panel detector has 2048 200 um square pixels.

Figure 5: Photograph of x-ray imaging equipment

The second x-ray imaging effort uses a much smaller flat panel direct digital detector, only 13 cm (5.1-inches) square. In the second effort, the pixels are 127 um square whereas in the first effort the pixels are 200 um square. The effort using the smaller detector requires many exposures to cover the entire 42 cm square (17-inch) armor panel while the effort using the larger detector are acquired in one exposure with little loss in spatial resolution.

Previous work at the facility using the larger detector[2] on use of x-ray imaging for ceramic materials suggested that detection sensitivity is impacted by the x-ray energy levels employed. Lower x-ray energies are preferred for higher resulting image contrast. In this previous work, two x-ray protocols were used in order to best establish defect detection. The two protocols employed are shown below in Table III.

Table III   X-Ray imaging protocols

Standard image acquisition (no saturation)	Low voltage image acquisition
150 kVp, 1.25 mA and 3.00, 1.5 mm spot size	70 kVp, 10.0 mA, 4.5 mm spot size
Tube to sample distance: 234.95 cm	Tube to sample distance: 234.95 cm
Beam filter: 0.127 mm copper	Beam filter: 1.6 mm aluminum
Integration Time: 570 ms	Integration Time: 2.280 Sec
Screen: none	Screen: none
100 frame average	50 frame average

In order to establish the detection sensitivity, the previous work utilized a line pair phantom made of aluminum and carbon rods. Shown below in Figure 6 is a typical through-transmission x-ray image of test panel A5 with the calibration phantom. These data were obtained at 150 kVp and at these x-ray energies the higher density center tile is not easily detected.

Figure 6 Calibration x-ray Image. Panel A5. Single center high density tile. 150 kVp data. Line pair phantom, Aluminum and carbon rod image quality indicators visible in upper center tile column.

3. Scanning Microwave Methods

Initially, plans called for exploring the use of the scanning microwave interference technique[3] where the interference pattern is created by irradiating the part in microwave energy as illustrated in Figure 7. The probe (transmitter and receiver antenna) is raster-scanned over the part and the signal at the receivers is sampled. The detected voltage values represent differences in the local dielectric constant. The voltage values for both receivers are saved with the associated X-Y position on the object. The saved voltages are displayed as a function of X-Y position on a computer monitor. The resulting "image" is displayed as a surface; with each X-Y position being shown as a gray scale, a false-color or as a 3D Z value of the part surface.

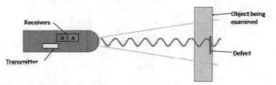

Figure 7 Schematic diagram showing relative position of microwave transmitter and receiver head to the part under examination. One-sided access is shown.

The scanning microwave method is best employed when the material to be examined is non-electrically conducting. In the case of these layered composite armor samples, certain materials were used that were electrically conducting and thus this NDE method is not applicable. It is to be noted that research is now being conducted to counteract this restriction for armor applications.

4. Immersion, Single-Transducer, Scanning Ultrasound

Immersion ultrasound[4,5] has long been used for defect detection in a wide variety of materials. There are two modes for this technique: through transmission, wherein one transducer, an emitter, is placed on one side of the object, and a second transducer, a receiver, is placed on the opposite side of the object. This works best if the specimen under study is a flat plate with nearly parallel sides. The second mode for immersion ultrasound is called pulse-echo. In this case a single transducer is used to both transmit and receive. Through use of special electronics and digital controls, there is a simple time delay between the transmitted signal and the echo—or received signal. The ultrasonic mechanical scanning tank and related hardware are similar to the phased-array scanning ultrasound system with the exception that the driving electronics for the transducer are simpler and the transducer itself is a single element transducer. In contrast to the phased-array method , the single element transducer method cannot be focused at different positions within the object unless the stand-off distance between the transducer and the test part is changed. Thus to scan with the "focus" of a single transducer, one must make several scans and between scans, the stand-off distance has to be changed. This usually results in very long data acquisition times as compared to phased-array scans.

5. Air Coupled Ultrasonic Methods

Air-coupled ultrasound[6,7] is a relatively new technology. This method eliminates the need for any liquid coupling between the test object and the ultrasonic probe. However, the method is limited to use in the through-transmission mode because of the low acoustic energy insertion. Further, because of issues with fabrication, usually air-coupled transducers are limited to frequencies less than 1 MHz. The limitations imposed by use of low frequencies thus also limit the defect size that can be detected. The authors have explored various air-coupled ultrasonic transducers including both piezo-electric and capacitance. Piezo–electric air-coupled transducers tend to be of high Q and thus offer little subsequent digital spectral analysis. Capacitance-based air-coupled transducers, while offering broad-band sound and thus the potential for digital spectral analysis, provide very low acoustic signals and thus offer lower signal to noise ratios. In the work for this project, air-coupled ultrasound methods using through-transmission with piezo-electric transducers are being explored by a second facility.

RESULTS

The results suggest that defect detection sensitivity for each NDE method has several dependency factors. In the case of ultrasound, both phased-array and single-transducer, the effects of the different materials in each layer impacts the way the elastic wave is focused and thus impacts the signal to noise ratio. Thus it is best to insonify from the side of the material where there are fewer interfaces. Shown below in Figure 8 are examples of water-immersion phased-array ultrasound used to detect the higher density ceramic tile in a complete layered ceramic-composite armor plate. It is also to be noted that the darker blue regions, within the yellow green regions, are thought to be regions of inconsistent adhesion but this has to be verified in future work. These plates were about 40 cm square and contained several layers of different materials. On the left it is shown that there is a single higher density tile and on the right it is shown that there are three higher density tiles.

(a)   (b)

Figure 8   Phased-array ultrasound data used to detect higher density ceramic tile in a complete ceramic-composite armor appliqué. a) Single higher density center tile, b) three clustered higher density tile. To be noted is that there are regions within the tile where inconsistent adhesion is suspected.

For x-ray projection direct-digital imaging, there is little if any effect on whether or not the ballistic impact side is towards the x-ray source or towards the x-ray detector. However, there is a significant detection sensitivity difference as a function of incident x-ray energy.   Figure 9a below shows a 150 kVp projection image of a test panel with a single higher density center tile and Figure 9b shows a similar confirmation test panel imaged at 70 kVp. Note that Figure 10b from the 70 kVp x-ray more readily detects the higher density center tile and is differentiated from the lower density outer tile. However, in no x-ray imaging tests were any suggestions of detection of the intentional delamination shown.

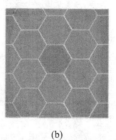

(a)   (b)

Figure 9   X-ray image of Panel C1. a)- 150 kVp projection image and b)-70 kVp projection data with gray scale adjusted

Based on the results of these initial tests, the following observations have been made:
1—Use of low kVp imaging techniques allows better contrast detection between tiles of different materials.
2—Use of simple digital image processing is necessary for better visualization of contrast differences.
3—Initial analysis x-ray image data suggests that while the suggested "delaminations" are not likely to be detected; the x-ray image data clearly demonstrates the ability for detection of small changes in the gap spacing among the tile.

As a comparison between single transducer ultrasound pulse-echo and phased array ultrasound, Figure 10 below shows test data from armor sample C5. Both data sets were acquired in the same manner—that is insonification from the low elastic modulus side. The advantage of the phased array image data is that there is a higher signal to noise ratio as observed in the larger image contrast. Such larger image contrast values corresponds to a better ability to detect small changes in the sample.

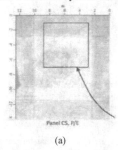 

(a)   (b)

Figure 10   Comparison between a) single transducer ultrasound data and b) phased array ultrasound data on same C5 test sample. Both immersion data sets acquired using same insonification side

No data were acquired with the scanning microwave system once it was discovered that there was an electrical conductor in the material set.

Data have also been acquired with air-coupled ultrasound. The transducers used in a through-transmission mode were 120 kHz focused transducers. Again, the effect of the layers is reduced because as with the x-rays systems, the data are integrated in the through-transmission mode . To date, the detection sensitivity shown by air-coupled ultrasound is far inferior to that of either of the the immersion ultrasound techniques or direct digital x-ray imaging.

CONCLUSIONS

In conclusion, as part of a larger multi-task Effects of Defects project to establish NDE techniques for detecting damage levels in ceramic-composite armor, 84 specially made ceramic-composite armor test panels have been produced.   Several forms of ultrasonic testing have been used: Immersion phased-array, immersion single-focused transducers, and air-coupled systems. The advantage of phased-array scanning is that it is faster, is able to focus at a depth dynamically and has better detection sensitivity. Two x-ray imaging systems have been explored: both using through-transmission, direct digital systems. One uses a single large-area flat panel amorphous silicon based detector whereas as the other uses a small-area, amorphous silicon, flat panel. The use of a single large-area flat panel is preferred because it significantly reduces time for data acquisition, hence cost, with little spatial resolution loss—certainly for the defects of this project. Scanning microwaves were initially explored but are presently unusable if an electrical conducting layer is employed in a layered armor structure. It seems that phased-array ultrasound and use of large-area, direct-digital projection x-ray imaging offers the most time and cost effective NDE technologies for armor **not** mounted on vehicles. The future work on this project, to be completed next year, will utilize digital image processing methods to analyze all test data acquired. Use of digital image processing methods will allow signal to noise ratios to be established from the various intentional defect panels. It is also planned that a limited amount of destructive analysis will be conducted of the intentional defect panels to verify exactly what the delaminations are like. This will allow correlations to be established between the NDE data and the defect characteristics.

REFERENCES

[1] J.S. Steckenrider, W.A. Ellingson, and J.M. Wheeler, Ultrasonic Techniques for Evaluation of SiC Armor Tile , *Ceramic Eng. And Sci. Proc.* 26(7) 215-222 (2005)

[2] W. A. Ellingson and M. W. Vannier , Applications of Dual-Energy X-Ray Computed Tomography to Structural Ceramics,in Proc. 37th Ann. Denver Conf. on *Advances in X-Ray Analysis*, , ed. C. S. Barrett, J. V. Gilfrich, R. Jenkins, T. C. Huang, and P. K. Predecki, Plenum Press, NY, 629-640 (1989)

[3] K. Schmidt, J. Little and W. Ellingson, a portable microwave scanning technique for nondestructive testing of multilayered dielectric materials", Proc. 32nd Inter. Conf. & Expos. on Adv. Ceramics and Composites, (2008)

[4] Y. Ikeda., K.Onda, and Y. Mizuta, Nondestructive Evaluation of Porosity in Si3N4 Ceramics by using Ultrasonic method, in *Proc. Of the 4th Symposium on Ultrasonic Testing, JSNDI*,105-108 (1997)

[5] Nondestructive testing handbook, 2nd ed., Vol 7—Ultrasonic Testing, P. McIntire, Ed. Am Soc. For Nondestructive testing (1991)

[6] T. A. K. Pillai, W. A. Ellingson, J. G. Sun, T. E. Easler , and A. Szweda, A Correlation of Air-Coupled Ultrasonic and Thermal Diffusivity Data for CFCC Materials, in *Ceramic Engr. and Sci. Proc.,* Vol. 18, Issue 4 (1997).

[7] W.A. Grandia and C. M. Fortunko, NDE applications of air-coupled ultrasonic transducers, in *Proc. IEEE Ultrasonic symposium,* Vol.1, 697-709(1995)

NDT CHARACTERIZATION OF BORON CARBIDE FOR BALLISTIC APPLICATIONS

Dimosthenis Liaptsis* and Ian Cooper
TWI NDT Validation Centre (Wales) Ltd, Port Talbot, Margam, SA13 2EZ, United Kingdom
Nick Ludford and Alec Gunner
TWI Ltd, Granta Park, Abington, CB21 6AL, United Kingdom
Mike Williams and David Willis
Kennametal Sintec, Newport, NP19 4SR, United Kingdom
Colin Roberson and Lucian Falticeanu,
Advanced Defense Materials Ltd, Rugby, Warwickshire, CV21 3XH, United Kingdom
Peter Brown
DSTL, Salisbury, Wiltshire, SP4 0JQ, United Kingdom

ABSTRACT
Boron Carbide (B_4C) is widely used to provide ballistic protection in many challenging service environments. This work was undertaken to determine a suitable process for NDT characterization of Boron Carbide. The project involves the introduction of deliberate flaws within the Boron Carbide tiles during manufacturing. An extensive experimental NDT program has been undertaken to determine the best method to detect defects into the ceramic tiles as well as to determine density variations that can be used for quality monitoring purposes. Ultrasonic characterization of the material was found to be able to detect both density variation and defects within the material. The defect resolution of the technique was found to be dependent upon the focus depth of the ultrasonic beam. The data obtained from the ultrasonic characterization was compared with other NDT techniques such as microwave and radiographic testing. This paper presents some of the data obtained during the NDT testing to demonstrate the capabilities and limitations of each of the techniques. A number of the supplied tiles were subjected to testing with all the three presented techniques as means of comparison of the different NDT techniques.

INTRODUCTION

Ceramic materials such as Boron Carbide (B_4C) and Silicon Carbide (SiC) are usually incorporated into armour systems in order to reduce their weight while providing high hardness, high compressive and tensile strength, and good response to elastic stress. However, the presence of critically sized defects and flaws in ceramic armour, such as pores and inclusions, or density variations across the tile area can compromise the ballistic performance and therefore lead to ballistic failure. An extensive NDT work has been carried out and published on SiC by a number of researchers[1-7]. However, a limited work can be found in the literature for the NDT testing of Boron Carbide ceramic material[8].

This paper presents the extensive NDT testing carried out on the supplied B_4C to determine the quality, as well as to detect defects that have been deliberately introduced into the ceramic tiles. For that purpose, a number of different NDT techniques have been assessed for their defect detection and sizing capabilities. The aim of this experimental program was to identify the most suitable NDT technique and to use this technique for further development and optimization.

Non-destructive testing methods such as ultrasonic immersion C-scan imaging provide a valuable method to examine and assess the bulk of the ceramic material. This technique can be used for both defect detection and density variation detection across the tile area. The ultrasonic immersion technique will be compared with radiographic and microwave testing.

EXPERIMENTAL PROCEDURE AND RESULTS

Boron Carbide tiles used in this study were manufactured and supplied by Kennametal Sintec (Newport, UK). The dimensions of the B_4C tiles supplied for NDT testing are 150mm x 150mm with a thickness of 10mm (except Pressing 7). The tiles in this experimental program were identified by number (#1 to #12) corresponding to their vertical position in the furnace during the pressing, i.e. Tile #1 was at the bottom and Tile #12 at the top. Both surfaces of each of the ceramic tiles were machined flat at the required thickness. Table I shows the tiles that were provided for testing. The manufacturing of the supplied tiles is ranging from using the standard process parameters to tiles pressed with different manufacturing parameters. Furthermore, a number of tiles had deliberately introduced inclusions that were used to assess the capabilities of the NDT techniques. For the introduction of deliberate defects in the volume of the ceramic tiles wood and graphite particles were used as well as graphite spheres. The size of particles is ranging from 0.2mm-1mm and the quantity used for each of the tiles differs. The aim was to determine the defect detection capabilities of the various techniques. Table I gives details for each of the pressing and the condition of each of the tiles that were tested.

Table I. Boron Carbide Tiles Supplied for NDT Testing

Pressing	Quantity of plates	Comments
1	8	Standard benchmark pressing
3	10	Plates 1 & 2 = 10mm thickness; Plates 3 & 4 = 5mm thickness Plates 5 & 6 = 15mm thickness; Plates 7 & 10 = 10mm thickness Plates 8 & 9 = 10mm thickness – Inclusion graphite paper fold
4	10	Plates 1 & 2 = 10mm thickness – Wood inclusions $1mm^3$ (Plate 1 = 5cc & Plate 2 = 10cc); Plates 3 & 4 = 10mm thickness – Graphite inclusions $1mm^3$ (Plate 3 = 5cc & Plate 4 = 10cc); Plates 5 & 6 = 10mm thickness – Graphite inclusions $1mm^3$ (Plate 5 = 15cc & Plate 6 = 20cc) Plate 7 = 10mm thickness – Graphite inclusions approximately $1mm^3$ (Plate 7 = 100cc); Plate 8 = 10mm thickness – Granulated nylon inclusions $0.5mm^3$ (Plate 8 = 5cc); Plates 9 & 10 = 10mm thickness – 8mm diameter holes (Plate 9 = 2.5mm deep & Plate 10 = 1mm deep)
5	12	Apply initial pressure ramp 100°C earlier than standard pressing
7	3	Plate 1 = 9mm thickness – 0.2mm-0.3mm graphite sphere inclusions Plate 5 = 9mm thickness – 0.4mm-0.5mm graphite sphere inclusions Plate 9 = 9mm thickness – 0.7mm-0.8mm graphite sphere inclusions

Ultrasonic Testing

In the immersion ultrasonic technique, the tiles were submerged into an X-Y scanning immersion tank. A beam of ultrasound is transmitted into the material, and the reflected energy is recorded and analyzed. The transducer is scanned across the tiles at increments of typically 0.5mm. A 7-axes ultrasonic immersion tank (USL7, Ultrasonic Sciences Limited, Aldershot, UK) used for the testing of the ceramic tiles that can operate at frequencies of up to 25MHz. The experimental set up used to carry out ultrasonic immersion scanning is shown in Figure 1. The focused immersion transducer was positioned at such a distance from the top or bottom surface of the tile as to focus on the material at depths of ¼ and ½ of the sample thickness. Using the experimental set up shown in Figure 1 the full volume of the tile was inspected with increased defect detection and sensitivity capabilities. Three ultrasonic scans were carried out for each of the ceramic tiles to cover the full tile volume at the highest detection sensitivity. Two of the scans were performed from the bottom surface of the tile and the third scan was carried out from the top surface, after the tile had been flipped over. This configuration was used as the types of defects found in ceramic materials are volumetric and are mainly inclusions, porosity and voids.

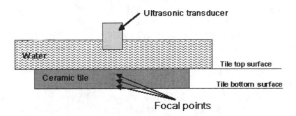

Figure 1. Diagram of the ultrasonic immersion testing experimental set up

The beam width is smaller at the focal point and, therefore, smaller defects can be detected. Also better sensitivity is achieved since the ultrasonic energy is concentrated at that point. In order to confirm the experimental parameters as well as to determine the focal spot size of the beam at the focal point, ultrasonic modeling was carried out using a specialized ultrasonic modeling software package (CIVA 9.2) developed from CEA (Commissariat a l'Energie Atomique). Ultrasonic modeling involves the definition of the material and ultrasonic properties of the component to be inspected, and the definition of the transducer parameters. The definition of these properties allows the ultrasonic modeling software to calculate the propagation of the beam within the material as well as the ultrasonic interaction and response from known defects. As an example, the probe was positioned at such a distance from the top surface of the ceramic tile to focus the ultrasonic beam at midwall of the ceramic material (i.e. at 5mm depth). Ultrasonic simulation was carried out to check that the transducer was positioned correctly to focus at different depths within the sample as well as to measure the beam spot size at the focal point. Figure 2a shows the beam propagation through the water and into the material, while Figure 2b shows the beam propagation and the ultrasonic focal point into the ceramic material. In that example, the depth where the maximum amplitude occurs (i.e. at the focal point) was measured at 5mm, which confirms the validity of the experimental set up used for the immersion scanning. At the focal point the ultrasonic energy is at maximum while the beam width is smaller. These focal point characteristics translate to better defect detection and response compared with an unfocused beam. The simulation shown that the -6dB beam width at the focal spot is 1mm.

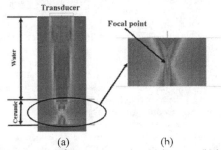

(a)   (h)

Figure 2. Visualization of the ultrasonic beam propagation (a) in water (b) in material

Pressing 1 Ultrasonic Results

For calibration and technique development purposes, Tile 1 from pressing 1 was machined with a diamond saw, and slots generated at different depths across the tile. Figure 3a shows a schematic diagram of the calibration tile used during the ultrasonic testing technique development. The slots have depths of 1mm, 2.5mm, 5mm and 7.5mm from the bottom surface. The tile was inspected from the top surface, and the purpose was to assess the penetration capabilities of the ultrasonic wave at various depths and different frequencies. Ultrasonic frequencies ranging from 5MHz to 50MHz (NOTE: Different immersion tank used for frequencies higher than 25MHz) were tested and it was concluded that the most suitable frequency for B$_4$C material was found to be the 15MHz transducer (Ultrasound Products, WC1003-04) with 76.2mm (3") focal length in water. All the slots and their depths in the tile can be clearly seen in the Time of Flight (ToF) data shown in Figure 3b. In ultrasonic immersion testing the ultrasonic transducer is scanned across the area of the tile. This allows the creation of two-dimensional C-scan maps corresponding to the ToF of ultrasound through the material. Images of the changes in ToF over each tile area were collected by gating the top and bottom surface reflected signals and determining the difference between them[2], or ToF, to obtain the transit time that in turn the software is translating into millimeters (mm) giving the tile thickness. The ultrasonic inspection system is monitoring the top and bottom surface reflected signals as well as the material volume to record the ToF data and the presence of indications in the tile. Using the ToF data in conjunction with the precise thickness of the tile, the longitudinal ultrasonic velocity can be accurately calculated as well as the depth of any indications detected within the tile. It should be noted that all the ultrasonic data presented in this paper was acquired using the 15MHz probe with 76.2mm (3") focal length in water.

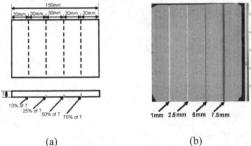

(a)   (b)

Figure 3. (a) Drawing of the calibration sample (b) Time of Flight scan data from the calibration sample

Furthermore, the ToF data can be used to obtain information regarding density variations across the tiles. Figure 4 shows the ToF data scans for tiles 1 and 2 from pressing 1. Here, the depth range of the ToF data presented in the scan has been changed from 0-11mm to 9.5-11mm. The aim was to detect changes in the material and, by using a shorter depth range; shuttle changes in the interaction of ultrasound with the ceramic material can be more easily seen. The color variations across the tile indicate variations in the density of the material. One of the limitations of the software used to capture the ultrasonic data is the inability to account for velocity changes in the sample due to changes in the material density. Thus, if there is considerable change in the velocity of the material, the software assumes there is a change in the thickness of the sample. As explained previously, the tiles have been ground to a uniform thickness, and any changes in the ToF data is attributed to velocity changes as a result of density changes. The ultrasonic velocity was set up for each of the tiles based on the tile thickness. The ultrasonic velocity for each tile was calculated by positioning the probe at a random point across the tile area, and readjusting the velocity value to get the correct value of the tile thickness. By using this velocity value the whole area of the tile was inspected. If there is a change in the material density, and therefore a change in the ultrasonic velocity, this will translate to color variations in the C-scan data. The color variations in tiles 1 and 2 are predominant indicating velocity changes, and therefore density changes. The probe was positioned to the different color regions in both tiles, and ultrasonic velocity measurements were taken across each of the regions. Figure 4 shows the measured velocity for each of the tiles across different color regions. It can be seen that tile 1 has two regions that are believed to have different material density. The velocity around the edge of the tile 1 was measured at 11,800m/s, while the velocity in the center of the tile was 12,200m/s. However, three density variation regions have been identified in tile 2 with measured velocities of 12,500m/s, 12,800m/s and 12,950m/s. Higher ultrasonic velocities indicate higher material density, while lower velocities indicate lower density.

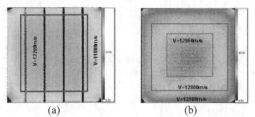

(a)   (b)

Figure 4. ToF scan data from pressing 1of (a) Tile 1 (b) Tile 2

The ToF scan data for the rest of the tiles from pressing 1 shown density uniformity. It was also noticed that each of the tiles exhibits a reduction in density around its edges and corners due to the pressure profile during the manufacturing.

Pressing 4 Ultrasonic Results

The manufacturing of pressing 4 tiles involved the deliberate introduction of flaws. Wood and graphite inclusions of approximately 1mm^3 were added into the B$_4$C powder before the pressing. Different quantities of inclusions were added to each of the tiles, as detailed in Table I. All the tiles from pressing 4 have a thickness of 10mm and were produced using the same manufacturing parameters as in pressing 3 (initial pressure ramp applied 100°C earlier). Focusing at different focal depths from both surfaces, all the B$_4$C tiles were inspected. It should be mentioned that the ultrasonic velocity was measured at 13,850m/s for all the tiles in pressing 4.

Figures 5 show the ToF data obtained from some of the pressing 4 tiles at different focal depths. It can be seen that the added inclusions in the tiles have been detected using the developed immersion testing. The flaw indications are shown in different colors that denote their depth in the tile. It should also be noted that the flaws are randomly distributed across the area and thickness of the tiles. From Figures 5 it can be seen that additional information can be obtained by focusing the ultrasonic beam at different depths. Indications that cannot be seen in one of the scans are detectable in a different scan. As an example, indications detected in Figure 5c ToF scan (shown in the blue circles) cannot be seen in the ToF scan in Figure 5b. This is also the case for the rest of the pressing 4 tiles inspected. This proves that the detection capability of the immersion testing is improved by focusing the beam at different depths across the tile thickness.

(a)   (b)   (c)

Figure 5. ToF data from Pressing 4 of tile 2 (a) Focus at 2.5mm from top surface (b) Focus at 2.5mm from bottom surface (c) Focus at 5mm from bottom surface

Figure 6 presents the amplitude data scan from two of pressing 4 tiles. The amplitude data from tiles 2 and 3 from pressing 4 is presented in order to demonstrate the sensitivity of the amplitude scan to the presence of inclusions in the ceramic tiles. The amplitude data presented shows the intensity of the signal reflected from the backwall of the tile. In summary, the data presented in the amplitude scan maps is explained by:

- Lighter grey corresponds to higher signal amplitude, i.e. lower attenuation.
- Darker grey corresponds to lower signal amplitude, i.e. higher attenuation.

In Figure 6a variations on the reflected signal amplitude were noted on the scan by red circles. Tile 2 from pressing 4 has 5cc of wood inclusions and because of the density difference between the wood and the ceramic material there are subtle differences in the material attenuation that has an effect to the beam propagation and consequently to the intensity of the reflected signal. Figure 6b presents the amplitude scan of tile 3 that has 5cc of graphite inclusions and the attenuation differences because of their higher density nature are more obvious compared to the wood inclusions.

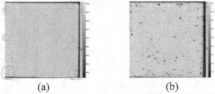

(a)   (b)

Figure 6. Amplitude data from Pressing 4 of (a) Tile 2 (b) Tile 3

ToF data for tile 2 from pressing 4 was presented in Figures 5 and although a number of inclusions deliberately introduced into the tiles were identified the amplitude scan was proven to be more sensitive to the density variations caused by the wood (Tile 2) and graphite (Tile 3) inclusions. The amplitude data was especially useful to detect surface or near surface inclusions introduced into the tile. However, the amplitude data can detect the presence of inclusions or voids into the tiles but cannot determine their depth. For the depth determination of a defect present in the tile the ToF data can be used.

Pressing 7 Ultrasonic Results

Pressing 7 tiles have graphite sphere inclusions supplied by Superior Graphite. The size of the spheres is ranging from 0.2mm to 0.8mm. Table 1 gives the details for each of the Pressing 7 tiles. Figure 7 shows the results from Tile 1 where 0.2-0.3mm graphite sphere inclusions have been introduced into the tile. Figure 7 shows the ToF data at focus depths of 4.5mm and 2.25mm as well as the amplitude scan data. It can be seen when the beam was focused at 4.5mm in the ceramic material a number of small inclusions have been detected across the tile area. The black area indicates that no ultrasound has been propagated and reflected at the backwall of the tile. This is believed to be caused due to the concentration of graphite inclusions in a small area of the tile that increased the material attenuation and therefore prevented the propagation of ultrasound into the material.

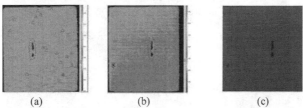

(a)   (b)   (c)

Figure 7. Tile 1 from pressing 7 (a) ToF data at focus depth of 4.5mm (b) ToF data at focus depth of 2.25mm (c) amplitude data

Similarly, tile 5 from pressing 7 has 0.4-0.5mm graphite spherical inclusions and was inspected ultrasonically. However it seems that that the graphite particles have not been properly distributed across the tile area but they appear as clusters. Figure 8 shows that in this case the ToF data cannot detect the graphite inclusions but variations in the amplitude data across the tile area indicates material attenuation differences, which in turn reveals the presence of higher density foreign bodies, i.e. graphite spheres.

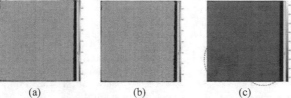

(a)   (b)   (c)

Figure 8. Tile 5 from pressing 7 (a) ToF data at focus depth of 4.5mm (b) ToF data at focus depth of 2.25mm (c) amplitude data

Tile 9 from pressing 7 has 0.7-0.8mm graphite spherical inclusions and was inspected ultrasonically. Figure 9 shows that in this case the ToF data has detect a number of the graphite inclusions but also variations in the amplitude data across the tile area indicates the presence of more graphite inclusions in the ceramic material.

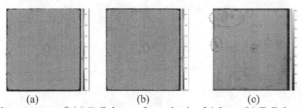

(a)   (b)   (c)

Figure 9. Tile 9 from pressing 7 (a) ToF data at focus depth of 4.5mm (b) ToF data at focus depth of 2.25mm (c) amplitude data

X-Ray Testing

One of the techniques assessed for the defect and density variation detection capabilities is digital X-ray radiographic testing. All the tiles that have been previously tested with ultrasonic immersion testing and presented in this paper were chosen for radiographic testing. The tiles were placed into the 225kV X-ray 5μm microfocus bay (X-Tek). The set up used to obtain the radiographic results is shown in Figure 10. The parameters used to obtain the radiographic images were optimized for defect detection. The X-ray energy used to penetrate and radiograph the B$_4$C tiles was 50kV.

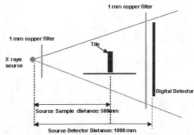

Figure 10. Radiographic testing experimental set up

Initially tiles 2 and 5 from pressing 1 were chosen to assess the sensitivity of X-ray testing to material density variations at the tiles. Tile 2 was previously tested with immersion testing and density variations were detected across the tile area (see Figure 4b). Tile 5 was also tested with the ultrasonic immersion technique and it was found that the material density is consistent across the tile area. Therefore, using these tiles the sensitivity and capability of the X-ray radiographic technique will be assessed. Figure 11 shows the radiographic images obtained from both B$_4$C ceramic armour tiles. As seen in Figure 11, both images are the same, and density variations found on tile 2 with ultrasonic testing cannot be seen across this tile. Also density differences between the two tiles are not noticeable at the radiographs. Note that the darker region at the bottom of the images does not correspond to material feature but is due to the polystyrene used to support the tiles during the radiographic testing.

(a)   (b)

Figure 11. Radiographic images from Pressing 1 of (a) Tile 2 (b) Tile 5

Figure 12 shows the radiographic images of tiles 2 and 3 from pressing 4. Tiles 2 and 3 had deliberately introduced 5cc of wood and graphite inclusions, respectively. It can be seen that the radiographic testing cannot detect the wood inclusions introduced into tile 2 but can detect some of the graphite inclusions (shown in blue circles in Figure 12b) in tile 3. These results indicate that the ultrasonic data for these tiles (Figures 5 and 6) detected more inclusions compared to the radiographic testing.

| (a) | (b) |

Figure 12. Radiographic images from Pressing 4 of (a) Tile 2 (b) Tile 3

Similarly, Figure 13 presents the radiographic images of tiles 1, 5 and 9 from pressing 7. Ultrasonic testing detected a number of inclusions in all these tiles and it can be observed that inclusions have not been detected or a small number of them detected by radiographic testing. Ultrasonic testing has proven to have increased defect detection and sensitivity capabilities compared to radiographic testing.

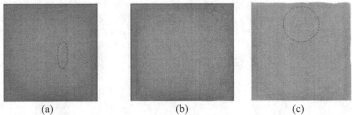

| (a) | (b) | (c) |

Figure 13. Radiographic images from Pressing 7 of (a) Tile 1 (b) Tile 5 (c) Tile 9

Microwave Testing

The Evisive Scan method uses a microwave transmitter to bathe the inspected component in microwave energy and uses multiple receivers to detect the integrated emitter and reflected signals. This method utilizes microwaves as an interrogating beam to penetrate a dielectric material. The microwaves are reflected at areas of changing dielectric constant. This reflection and the interrogating beam combine to form an interference pattern. The receiver measures the microwave signal as voltage differences at locations over the surface of the material that indicates the presence of a potential defect, or change in material dielectric properties[10]. The Evisive Scan™ equipment used for the microwave testing of the ceramic tiles consists of a microwave probe, process instrumentation, an interface computer and display and a probe positioning equipment that uses an OptiTrack IR (InfraRed) camera to follow the probe movement.

Figure 14 shows the microwave scan performed on the calibration tile with the machined slots used for the development of the ultrasonic technique. This tile was used to asses the penetration capabilities of microwave testing. The microwave scan depicts voltage variations in the ceramic material and it can be seen that all the machined slots cannot be detected since the presence of the slots will cause a voltage variations. This data shows that microwave testing is not a suitable technique to assess the B_4C tiles used for testing in this project because of the high material conductivity that limits the penetration capabilities of microwaves.

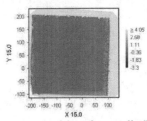

Figure 14. Microwave scan of the reference tile (Tile 1, Pressing 1)

DISCUSSION

Figure 15 shows the ultrasonic velocities measured from all the ceramic tiles tested. The changes in velocities in pressing 1 tiles indicate changes in density that is dependent upon the vertical position of the tile in the furnace during the pressing. From the ultrasonic results of pressing 1, it is evident that the ultrasonic velocity increases as a function of the vertical position of the tiles in the pressing. Lower ultrasonic velocities were measured in the tiles that were at the bottom in the furnace, while higher velocities exhibited from the tiles that are higher in the furnace during the pressing. The ultrasonic velocities at the centre of the tiles range from 12,200m/s for tile 1 to 13,450m/s for tile 10.

A difference observed in the measured ultrasonic velocities from pressing 3 tiles is the higher velocities exhibited compared with the values obtained from pressing 1 tiles. Even for the tiles placed at the lower part of the furnace higher ultrasonic velocities were obtained. As an example, tile 1 in pressing 1 has a velocity of 12,200m/s compared to tile 1 in pressing 3 with a velocity of 13,600m/s. In general, the velocities measured in the pressing 3 tiles range from 13,600m/s to 13,850m/s. Also, the ultrasonic velocity was uniform across the area of each tile, which indicates the material density uniformity and manufacturing quality of the ceramic armour tiles. Furthermore, the velocity and therefore the density of the tiles ware independent of their vertical position in the furnace during the pressing.

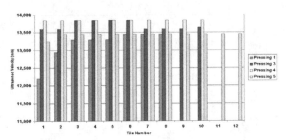

Figure 15. Ultrasonic velocity measurements from the B_4C tiles

The differences in the density quality of the pressings 3-7 tiles are attributed to the alteration of the manufacturing parameters during the pressing where the initial pressure ramp was applied 100° C earlier. In order to confirm the correlation between the ultrasonic velocity and the material density a number of tiles were sectioned and their density was measured. Density measurements were taken mainly from the centre of the pressed tiles in order to avoid any edge effects within the tiles.

Figure 16 shows the density measured from the sectioned tiles. As it can be seen from Figure 16, there is a good correlation between the ultrasonic velocity measurements and the measured material density. These results shown that the ultrasonic technique is able to distinguish small density variations of up to $0.01 g/cm^3$, which proves the sensitivity of the ultrasonic technique to detect subtle material density variations across the ceramic tiles.

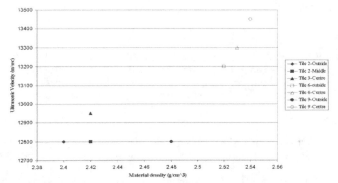

Figure 16. Correlation between material density and ultrasonic velocity

Radiographic testing was proven not to be as sensitive as the ultrasonic testing. Inclusions detected ultrasonically were not detected with the X-ray testing and whenever inclusions were detected with X-ray testing, more inclusions were found from ultrasonic testing by a combination of ToF and amplitude data. Additionally, the radiographic testing does not have very good defect detections capabilities close to the edge of the tiles. Radiographic testing was also proven not to be sensitive to density variations, which is contrary to the capabilities of the ultrasonic testing, where subtle material density changes at the tiles were detected.

Microwave testing was proven not to be a suitable technique for the characterization of the ceramic tiles tested in this project due to the high conductivity of the material that limits the capabilities of the technique.

CONCLUSIONS

Three different NDT techniques were assessed for their capabilities in detecting defects and density variations in B_4C ceramic armour tiles. Immersion testing was proven to be sensitive to material density variations, which is not the case with X-ray radiography and defect detection.

Deliberate defects were introduced into the boron carbide tiles using graphite and wood inclusions of approximately 0.2mm to 1mm in size. Immersion testing proved that these inclusions could be detected. The combination of ToF and amplitude data can provide information about the presence of defects, their size and depth as well as the presence of density variations across the ceramic armour tile. The immersion testing can also provide a measurement of the ultrasonic velocity of the material that is closely linked with the material density. A number of tiles that ultrasonic velocity variations detected were selected for sectioning and measurement of their actual densities. The sectioning and density measurements of the tiles demonstrated a good correlation between the material density and ultrasonic velocity.

Radiographic testing is not sensitive to detect material density variations and under certain circumstances can detect flaws in the material. The flaws detected from the radiographic testing were also detected from the ultrasonic testing but the ultrasonic testing was more sensitive to the detection of flaws embedded into the tile and present at the surface of the tiles. X-ray radiography testing is suitable for defect detection under specific conditions, such as where the defect is denser than the background material.

Microwave testing was proven not to be a suitable technique for the characterization of the ceramic tiles tested in this project due to the high conductivity of the material that limits the penetration capabilities of the technique.

ACKNOWLEDGEMENTS

This work was funded from Technology Strategy Board (TSB), UK. The authors would like to thank Peter Wikingsson from Superior Graphite for the provision of the graphite spheres used for the introduction of inclusions into the ceramic tiles. The authors would like also to acknowledge the contribution and support of Evisive Inc to the microwave testing of the ceramic tiles.

REFERENCES

[1]R. Brennan, R. A. Haber, D. Niesz, J. McCauley and M. Bhardwaj, Non-Destructive Evaluation (NDE) of Ceramic Armour: Fundamentals, Ceramic Engineering and Science Proceedings, 223-230, (2005).
[2]R. Brennan, R. Haber, and D. Niesz, Ultrasonic Evaluation of High-Density Silicon Carbide Ceramics, *Int. J. Appl. Ceram. Techol.*, **5 [2]**, 210-218 (2008).
[3]D. M. Slusark, M. V. Demirbas, A. Portune, S. Miller, R. A. Haber, R. Brennan, W. Green, E. Chin, and J. Campbell, Nondestructive Evaluation (NDE) of Sintered Silicon Carbide and Its Correlation to Microstructure and Mechanical Properties, 12th Annual Conference of Composites and Advanced Ceramic Materials, part 2 of 2, Ceramic and Science Proceedings, **9**, Issue 9/10 (2008).
[4]K. Lee, P. Lee, S. Cho, H. Lee, and A. Kohyama, Characteristic Evaluation of Liquid Phase-Sintered SiC Materials by a Nondestructive Technique, Journal of Nuclear Materials, **386-388**, 487-490 (2009)
[5]J. S. Steckenrider, W. A. Ellingson, R. Lipanovich, J. Wheeler, and C Deemer, Evaluation of SiC Armour Tile Using Ultrasonic Techniques, Advances in Ceramic Armour II, Ceramic and Science Proceedings, **27**, 33-41(2008).
[6]M. F. Zawrah and M. El-Gazery, Mechanical Properties of SiC Ceramics by Ultrasonic Non-Destructive Technique and its Bioactivity, **106**, 330-337 (2007).
[7]Y. M. Kim and E. C. Johnson, Determining the Material Properties of SiC Plates using NDE Methods, The Sixth International Conference of Condition Monitoring and Machinery Failure Prevention Technologies, 620-629 (2009).
[8]I. K. Sarpun, V Ozkan, S. Tuncel and R. Unal, Determination of Mean Grain Size by Ultrasonic Methods of Tungsten Carbide and Boron Carbide Composites Sintered at Various Temperatures, 4th International Conference on NDT, Hellenic Society for NDT (2007).
[9]G öller A, Microwave Scanning Technology for Material Testing, European Conference for NDT, 1-7 (2006).
[10]K. Schmidt, J. Little, and W. A. Ellingson, A portable microwave scanning technique for non-destructive testing of multilayered dielectric materials, Advances in Ceramic Armour IV, 179-189 (2009).

ADVANCED NONDESTRUCTIVE ULTRASOUND CHARACTERIZATION OF TRANSPARENT SPINEL

A. R. Portune and R. A. Haber
Rutgers University Materials Science and Engineering
Piscataway, New Jersey, USA

ABSTRACT

Materials chosen for lightweight transparent armor must demonstrate appropriate optical and mechanical properties, including high transmission across a wide range of the electromagnetic spectrum, high hardness, high strength, and good ballistic performance. Polycrystalline isotropic magnesium aluminate spinel ($MgAl_2O_4$) has long been considered ideal for this application due to its transparency in the 0.2µm – 5.5µm range and considerable mechanical properties. The presence of defects or precipitates in the finished spinel piece degrades both the optical and mechanical properties of the material. In this study advanced nondestructive ultrasound characterization has been performed on spinel materials with 0.6µm and 1.5µm mean grain sizes. C-scan mapping of acoustic attenuation reveals the location of large clusters of precipitate within the transparent ceramic. Acoustic spectroscopy performed in the 10 – 30MHz range characterizes the absorption of ultrasound by heterogeneities within the spinel microstructure. Analysis of ultrasound attenuation coefficient spectra provides insight into the relative concentration and mean size of clustered precipitate present within the bulk material.

INTRODUCTION

Transparent armor ceramics need to exhibit both excellent optical properties and adequate ballistic performance to be successful in their application. Candidates for transparent armor have traditionally been limited to various forms of fine grained polycrystalline ceramics or glass.[1] According to the Hall-Petch relationship, the materials' strength and hardness should increase as the grain size is reduced, translating to a general rise in the ballistic properties of the material. Leading candidates in transparent armor need to exhibit transparency into the infrared region of the electromagnetic spectrum in addition to high strength and hardness.[2] IR transparency is critical for many armor applications including infrared missile domes. Materials currently under consideration for these applications include AlON, sapphire, edge-form-growth sapphire, and magnesium aluminate spinel.[1]

Magnesium aluminate spinel ($MgAl_2O_4$) is a leading candidate for lightweight transparent armor due to its superior optical and mechanical properties.[3] Spinel has been developed since the early 1960's to enhance its intrinsic transparency and strength through advanced processing techniques and control of powder purity.[4-5] With porosity under 50nm, this material has shown transparency from the ultraviolet to the mid-IR in the 0.2µm – 5.5µm range.[2] Hot pressing and controlled heat treatments can achieve exceptionally fine microstructures with mean grain size as low as 0.6µm. Novel processing methods such as spark plasma sintering are also being investigated to achieve the best mechanical strength and hardness without sacrificing optical transparency.[6]

Sintering additives are added to spinel to control kinetic processes during formation of the material and to reduce undesirable impurities. The most common additive used to achieve low heterogeneity concentration is LiF.[3] Addition of 1.0% mass LiF enhances late-stage sintering of spinel through the formation of oxygen vacancies.[3] This additive also promotes the coalescence of pores into cylindrical cavities which sit along the grain boundaries where they are less detrimental to optical transparency.[2] The precipitation or clustering of sintering additives can cause spinel to appear hazy or foggy instead of transparent.[5] The compositional homogeneity of this material is critical to achieving its superior optical and mechanical properties.

Nondestructive characterization of microstructural uniformity can be undertaken using ultrasound techniques.[7] Ultrasound introduces oscillating elastic waves into the material which interact with microstructural features as they propagate. Measurements of the amplitude and time of flight (TOF) of reflected signal peaks can be used to calculate volumetric properties of the material over large sample areas.[8] Elastic properties such as Young's modulus and the longitudinal speed of sound are determined from TOF measurements of longitudinal and shear wave reflections. Attenuation of acoustic energy is determined from the reduction in peak amplitude for successive back surface reflections. Previous studies have shown attenuation measurements to be more sensitive to minute changes in composition compared with elastic property measurements.[9] Acoustic spectroscopy, measurement of the frequency dependency of the attenuation coefficient, has been shown to correlate with minute microstructural variations in solid materials.[10]

Acoustic Spectroscopy Theory

Acoustic spectroscopy has been used to correlate ultrasound results with microstructural features since early testing in metals was conducted in the 1950's.[11] By performing a fast Fourier transform (FFT) on back surface reflections, it is possible to measure the strength of individual frequencies in each signal. The Beer-Lambert law is then used to calculate the attenuation coefficient at each frequency, using the first bottom surface reflection as the initial intensity and the second bottom surface reflection as the intensity after propagation. This law is stated as:

$$I = I_0\, e^{-\alpha x} \tag{1}$$

If testing is performed in pulse-echo configuration, the path length is defined as twice the thickness of the sample. Figure 1 illustrates an oscilloscope view of labeled peak reflections showing the width of the time window used when performing the FFT on reflected signals. Note that the waveform in Figure 1 is meant for instructive purposes only, and does not represent peak reflections from the spinel samples examined in this study.

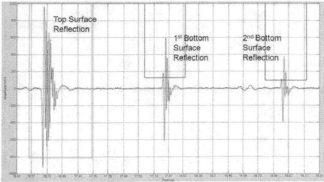

Figure 1: An oscilloscope view illustrating top and bottom surface reflections. The gates surrounding the bottom surface reflections correlate with the window used to perform the FFT on the reflected signal.

The attenuation coefficient measures the loss resulting from interactions with the bulk microstructure. Two mechanisms are responsible for energy loss in propagating waves – scattering and absorption. Scattering follows either Rayleigh or stochastic laws depending on the relationship between the size of the scatterer and the acoustic wavelength. For fine grained spinel, acoustic frequencies in excess of 100MHz would be necessary to render the Rayleigh approximation inadequate. Scattering in the Rayleigh range can be expressed as:

$$\alpha_{Grain Scattering} = A\, D^3 f^4 \qquad\qquad (2)^{12}$$

Where A is a constant, D is the grain diameter, and f is the ultrasound frequency. Significant changes in mean grain size throughout a sample area will result in large attenuation coefficient variations due to the cubic dependency on grain diameter.

Absorption of ultrasonic energy occurs from multiple sources, some intrinsic to the material and some reliant on the presence of heterogeneous phases. Intrinsic absorption occurs as a fraction of the ordered particle motion in the acoustic wave converts to disordered motion, or heat.[13] Extrinsic absorption has been best described in colloidal systems which are analogous to heterogeneous solid materials. Thermoelastic absorption occurs at thermal or elastic property interfaces as a result of field continuity.[14] Work in colloids has shown absorption phenomenon to dominate attenuation at lower frequencies where contributions from scattering are roughly negligible.[15] Absorption curves can be modeled as roughly Gaussian, whose behavior is defined as:

$$A\, e^{-\frac{(x-B)^2}{2C}} \qquad\qquad (3)$$

Where A defines the amplitude of the curve, B defines the central position and C defines the full width at half maximum. These can be correlated with the size distribution and concentration of heterogeneities within the bulk microstructure. If analogous to colloids, the A parameter would relate to the concentration of heterogeneities, the B parameter would relate to the mean size of the heterogeneity as well as its thermal and elastic mismatch with the bulk material, and the C parameter would relate to the width of the size distribution of heterogeneities within the volume of the bulk which the acoustic wave interacted with.[15]

EXPERIMENTAL PROCEDURE

Two magnesium aluminate spinel samples with 0.6μm and 1.5μm mean grain sizes were tested using advanced nondestructive ultrasound methods. The thickness of each spinel tile was measured to be approximately 12mm by taking the average of 10 digital caliper readings. The mean grain size value for each sample was provided by the manufacturer. Due to the proprietary nature of the microstructure, independent confirmation of these values was not attempted. C-scan mapping using a 20MHz central frequency planar transducer produced images of attenuation coefficient variations throughout the material. Figure 2 shows an FFT of the transducer output, demonstrating its -6dB bandwidth to be between 16-32MHz. High signal strength suitable for acoustic spectroscopy is achieved in the 10-34MHz range. C-scan maps found several large variations which correlated with yellow discolorations seen visibly in the samples. These variations were further examined using acoustic spectroscopy in the 10-30MHz range. Due to the grain size of the samples, acoustic spectroscopy results would be entirely in the absorption regime. Attenuation coefficient spectra were curve fit to a single Gaussian to determine the A, B, and C parameters of the curve. Analysis of variations within these parameters indicated relative changes in the concentration, mean size, and size distributions of precipitates within the spinel microstructure. Due to the proprietary nature of these samples, mechanical testing was not possible.

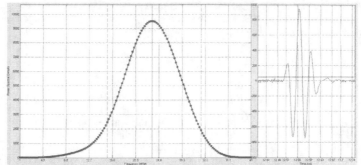

Figure 2: The signal output of the 20MHz planar transducer used in this study. The FFT taken was of the top surface reflection of a polished SiC mirror.

RESULTS

A C-scan map of the overall signal attenuation coefficient for the 1.5μm grain size spinel sample is shown in Figure 2. While most of the material displayed homogeneous results, several circular regions were seen with significantly higher attenuation coefficients. Three of these were more closely examined using acoustic spectroscopy, noted by numbers 1-3 in Figure 2. Region 1 centers on a large deviation from the mean behavior, region 2 centers on a small deviation from mean, and region 3 represents the mean acoustic response of the material.

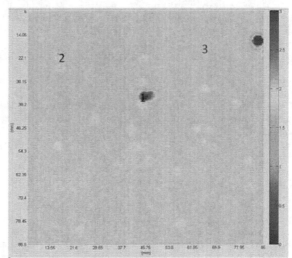

Figure 2: The C-scan map of overall signal attenuation coefficient measured using a 20MHz transducer on the spinel sample with 1.5μm mean grain size. The scale used is in units of dB/cm.

Acoustic spectroscopy results for the three chosen regions of the 1.5μm grain size spinel are shown in Figure 3. The approximately Gaussian shape of the measured curves confirms that the dominating attenuation mechanism in the frequency range examined is absorption. The results of curve fitting a single Gaussian to each attenuation coefficient spectra can be seen visually in Figure 3 as the dashed lines. Table I contains statistical information on the A, B, C, and R^2 values for each curve. A single Gaussian fits these attenuation coefficient spectra quite well, with the lowest R^2 value greater than 0.94. Analyzing changes in the Gaussian coefficients provides insight into changes in the microstructural homogeneity between different regions of the material.

Regions 1 and 2 show a significantly higher A coefficient relative to Region 3, which is used as a baseline for the average microstructure. This indicates that the concentration of heterogeneities such as precipitates is increased in these regions. The B coefficient for Region 1 has been shifted towards higher frequencies, indicating that the mean size of heterogeneities is likely smaller compared with the average microstructure. However, the C coefficient for this region is also significantly higher, which could indicate a wider size distribution of heterogeneities present within the bulk microstructure at this location. Similar behavior is seen for Region 2 although to a lesser extent. Acoustic spectroscopy results suggest that the regions of higher attenuation seen in Figure 2 are therefore due to a clustering of smaller heterogeneities rather than from large anomalous inclusions.

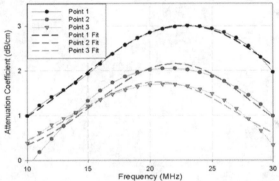

Figure 3: Acoustic spectroscopy results for three regions in the 1.5μm grain size spinel single Gaussian curve fitting.

Table I: Curve fitting parameters for attenuation coefficient spectra from the 1.5μm grain size spinel.

	Point 1	Point 2	Point 3
A	3.010	2.163	1.751
B	22.78	21.97	20.56
C	12.05	8.835	9.099
R^2	0.9945	0.9412	0.9651

A C-scan map of the overall signal attenuation coefficient for the 0.6μm grain size spinel sample is shown in Figure 4. Three regions were chosen for this material for more detailed analysis using acoustic spectroscopy, denoted by the numbers 1-3 in Figure 4. This material displayed far greater homogeneity in acoustic response relative to the 1.5μm grain size material. Only one highly attenuation region was located, denoted as Region 1. A second region which was slightly higher than the mean response was chosen as Region 2. Region 3 represents the mean microstructure of the material, and can be viewed as a baseline for comparing results. Note that a slightly wider scale is used in Figure 4 relative to Figure 2. This accounts for the higher attenuation coefficient values seen in Region 1 due to the large and visible precipitate located there.

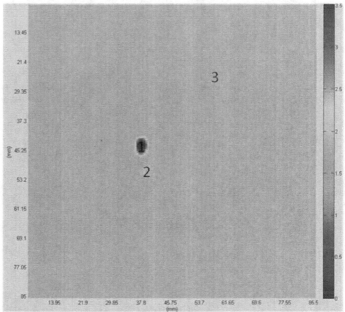

Figure 4: The C-scan map of overall signal attenuation coefficient measured using a 20MHz transducer on the spinel sample with 0.6μm mean grain size. The scale used is in units of dB/cm.

Acoustic spectroscopy results for the three chosen regions of the 0.6μm grain size spinel are shown in Figure 5. Like the 1.5μm grain size spinel, attenuation spectra from this material are approximately Gaussian in shape. The results of curve fitting a single Gaussian to each attenuation coefficient spectra can be seen visually in Figure 5 as the dashed lines. Table II contains statistical information on the A, B, C, and R^2 values for each curve. These curve fits produce R^2 values greater than 0.97, indicating that a single Gaussian is an appropriate choice for describing curve behavior. An analysis of changes in the Gaussian parameters provides insight into microstructural changes between regions of this material. The vertical scale used in Figure 5 is wider than that used for the previous sample in Figure 3 to account for the higher attenuation coefficient values measured.

Region 1 shows a significant increase in the A parameter relative to the other two regions of this material, indicating a marked rise in the concentration of heterogeneities present within this part of the microstructure. The B parameter shows a shift towards higher frequencies relative to the other two regions, indicating that the mean size of heterogeneities is likely smaller relative to the mean microstructure of the material. The C parameter for region 1 is almost double that of other regions, indicating that the width of the size distribution of heterogeneities is also increased for this part of the material. Region 2 shows very similar behavior to the baseline Region 3 except for a small increase in the A parameter. This indicates a minor increase in the concentration of heterogeneities present within the bulk without a significant change in their mean size or size distribution.

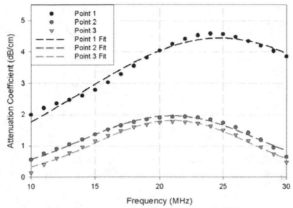

Figure 5: The attenuation coefficient spectra from the 0.6μm grain size spinel with single Gaussian curve fitting.

Table II: Curve fitting parameters for attenuation coefficient spectra from the 0.6μm grain size spinel.

	Point 1	Point 2	Point 3
A	4.446	1.968	1.820
B	24.76	20.95	21.05
C	15.30	9.802	8.458
R^2	0.9786	0.9814	0.9841

CONCLUSIONS

The Gaussian nature of frequency dependent attenuation coefficient measurements confirms that absorption is the dominating loss mechanism in both samples. This is unsurprising since the average grain size is two orders of magnitude smaller than the acoustic wavelength in the 10-30MHz range. Acoustic spectroscopy results from both materials show that clustering of small heterogeneities is typically the cause of high attenuation coefficient regions seen in C-scan maps. Fewer clusters were seen in the 0.6μm grain size spinel relative to the 1.5μm sample. As the magnitude of attenuation ranks with the concentration of bulk features, it is likely that the cluster seen in the 0.6μm grain size spinel contained a higher concentration of precipitates when compared to features seen in the 1.5μm material. Correlating acoustic results with FESEM images or chemical analysis was not possible in this study due to the proprietary nature of the materials investigated.

While acoustic spectroscopy can measure changes in the relative concentration and size distribution of these features, it is unable to make specific quantitative predictions at this time. Due to the nature of the experimental process, this technique is only suitable for materials with relatively smooth and parallel surfaces. Roughness becomes increasingly deleterious to attenuation coefficient measurements as frequency increases. Acoustic spectroscopy is applicable to all ceramic materials, whether single crystal or polycrystalline. As active loss mechanisms will vary from one material to the next, accurate interpretation of spectra results requires some a priori knowledge of the system under interrogation.

Advanced ultrasound characterization techniques have been demonstrated as successful in rapidly identifying microstructural variations within transparent armor ceramics. Comparing acoustic spectroscopy results from multiple sample regions enables an understanding of how the concentration and size distribution of heterogeneities within the bulk microstructure varies within the material. Gaussian curve fitting supplies three coefficients which correlate with changes in the size distribution and concentrations of heterogeneities within the bulk microstructure. A thorough study incorporating mechanical sectioning and imaging would enable more specific predictions about these microstructural factors. Since clusters of elastic heterogeneities could potentially reduce the mechanical and optical properties of transparent armor ceramics, determining the location and concentration of these microstructural features is crucial to accurate predictions of material performance in its application.

ACKNOWLEDGEMENTS

The authors would like to thank the NSF IUCRC Ceramic and Composite Materials Center and the Army Research Laboratory Materials Center of Excellence for Lightweight Vehicular Armor.

REFERENCES
[1] C. G. Fountzoulas, J. M. Sands, G. A. Gilde, and P. J. Patel, Modeling of Defects in Transparent Ceramics for Improving Armor, *Proceedings of the 24th International Symposium on Ballistics,* Vol. 2, pp. 760–767 (2008).
[2] I.E. Riemanis, H. Kleebe, R. Cook, and A. DiGiovanni, Transparent Spinel Fabricated from Novel Powders: Synthesis, Microstructure, and Optical Properties, *10th DoD Electromagnetic Windows Symposium,* (2004).
[3] K. Rozenburg, I. Riemanis, H. Kleebe, and R. Cook, Sintering Kinetics of a $MgAl_2O_4$ Spinel Doped with LiF, *J. Am. Ceram. Soc.,* **91** [2], 444–450 (2008).
[4] H. Palmour, Development of Polycrystalline Spinel for Transparent Armor Applications, Final Technical Report, Army Materials and Mechanics Research Center, Massachusetts (1972).
[5] R. Cook, M. Kochis, I. Riemanis, and H. Kleebe, A new powder production route for transparent spinel windows: powder synthesis and window properties, *Proceeding of the Defense and Security Symposium* (2005).
[6] N. Frage, S. Cohen, S. Meir, S. Kalabukhov, and M.P. Dariel, Spark plasma sintering (SPS) of transparent magnesium-aluminate spinel, *J. Mater. Sci.,* **42,** 3273–3275 (2007).
[7] R. Brennan, Ph.D. Thesis Dissertation, Ultrasonic Nondestructive Evaluation of Armor Ceramics, Rutgers University (2007).
[8] R. Brennan, R. Haber, D. Niesz, G. Sigel, and J. McCauley, Elastic Property Mapping Using Ultrasonic, *Ceramic Engineering and Science Proceedings,* **28** no. 5, 213-222 (2008).
[9] S. Bottiglieri and R. Haber, High Frequency Ultrasound of Armor-Grade Alumina Ceramics, *Proceedings of the 35th Annual Review of Progress in Quantitative Nondestructive Evaluation,* Volume **1096,** 1301-1308 (2009).
[10] A. Vary, and H. Kautz, Transfer Function Concept for Ultrasonic Characterization of Material Microstructures, Materials Analysis by Ultrasonics – Metals, Ceramics, Composites, Noyes Data Corporation, Park Ridge, New Jersey, 249-289 (1987).
[11] W. Mason and H. McSkimin, Attenuation and Scattering of High Frequency Sound Waves in Metals and Glasses, *The Journal of the Acoustical Society of America,* **19**(3), 464-473 (1947).
[12] D. Nicoletti, N. Bilgutay, B. Onaral, *Ultrasonics Symposium Proceedings,* 1119 – 1122 (1990).

[13] L. Luo and J. Molnar, Ultrasound absorption and entropy production in biological tissue: a novel approach to anticancer therapy, *Diagnostic Pathology,* **1**(35), (2006).
[14] K. Lucke, Ultrasonic Attenuation Caused by Thermoelastic Heat Flow, *The Journal of Applied Physics*, **27**(12), 1433-1438 (1956).
[15] A. Dukhin, and P. Goetz, Ultrasound for Characterizing Colloids – Particle Sizing, Zeta Potential Energy, Elsevier Science B. V., Amsterdam, Netherlands, 2002.

OPTIMIZATION OF A PORTABLE MICROWAVE INTERFERENCE SCANNING SYSTEM FOR NONDESTRUCTIVE TESTING OF MULTI-LAYERED DIELECTRIC MATERIALS

K. F. Schmidt, Jr. J. R. Little, Jr.
Evisive, Inc.
Baton Rouge, Louisiana USA

W. A. Ellingson
Argonne National Laboratory
Argonne, Illinois USA

L. P. Franks
US Army Research and Development Command Tank Automotive Research Development and Engineering Center
Warren, Michigan, USA

W. Green
US Army Research Laboratory
Aberdeen Proving Ground, Maryland, USA

ABSTRACT

A portable microwave interference scanning system using the Evisive Scan™ microwave interference scanning method has been demonstrated to detect damage in composite ceramic armor test specimens including engineered features in specially fabricated surrogates. The antenna system has been optimized for detection of identified features in sample specimens. The system configuration; hardware, firmware, software and communications protocols have been optimized for portability and field application flexibility. The optimization has included wireless interfacing of the antenna, system electronics and computer control.

The microwave interference scanning technique detects and images internal cracks, internal laminar features such as disbonds and variations in material properties such as density. It requires access to only one surface, and no coupling medium. Other NDE methods, including through-transmission x-ray, x-ray Computed Tomography, and destructive examination, are used to verify defects and other detected anomalies. The development of this portable system will provide a suitable method for in-theatre health monitoring of composite ceramic armor. Work has shown that detected damage level data are not affected by outer covering layers. Test panels used in this work were provided by the US Army Tank-Automotive Research, Development and Engineering Center (TARDEC), by the US Army Research Laboratory, and by the Ballistics Testing Station through Argonne National Laboratory. This paper will describe the system and present current results. This work is supported by US Army *Tank-Automotive Research, Development and Engineering Center (TARDEC) and US Army Research Laboratory*.

INTRODUCTION

The Evisive microwave interference scanning technique has been successfully demonstrated on armor panels constructed of high-performance technical ceramics. The ceramic armor is employed in the form of plate inserts in garments and seats; in panels in vehicles, aircraft and vessels; and as an appliqué in armored vehicles. Ceramic armor provides effective and efficient erosion of and defeat of ballistic threats. Effectiveness of ceramic armor can be degraded by defects present from production and by operational damage resulting from handling or impact with objects in the environment, other than projectiles. In normal use, ceramic armor is routinely exposed to the possibility of such damage[1].

A means to detect damage and manufacturing defects which are not visually apparent is needed to determine the integrity of the ceramic armor so that appropriate replacement can be made. Recently, a microwave-based method, having US and international patents [2,3,4,5,6,7], has been developed and demonstrated that is as applicable to ceramic armor systems (reported here in 2008[8] and 2009[9]). Applications development has included optimization of antenna – material interaction, and miniaturization of the Evisive Scan[TM] system. The Evisive Scan method permits real time evaluation by inspection from one surface only, through non-contacting encapsulation, with panels hung in place.

DESCRIPTION OF THE METHOD

Evisive Scan method requires access to only one side of a part. The microwave interference pattern is created by bathing the part in microwave energy as illustrated in Figure 1. The probe (transmitter and receiver antenna) is moved over the part, bathing it in microwave energy. Some energy is reflected and transmitted at every interface of changing dielectric constant in the field of the transmitter. This includes the front and back surfaces of the part, and every "feature" in the part that has a discontinuity in dielectric properties. A microwave interference pattern is created when the reflected energy is combined with the transmitted signal to create the measured detector voltage at each

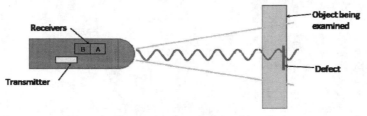

Figure 1.   Schematic diagram showing relative position of microwave transmitter and receiver head to the part under examination. One-sided access is shown.

of the receivers. The voltage values for both receivers are saved with the associated X-Y position on the object.

The combination of dielectric constants of the engineered ceramic materials and the microwave frequency used in the tests yields wavelengths in the material of about 8 mm (0.33 inches) to 20 mm (0.83 inches). The magnitude of the phase difference between the emitted signal and reflected signal determines the voltage of the signal. This interference pattern is illustrated by the response from a 2.5 mm (0.10 inch) spherical conductive reflector shown in Figure 2.

Hardware Channels A and B are separated by a quarter wavelength ($\lambda/4$) in the wave propagation dimension, Z. In any "image" data, the rate of change of the detected signal value impacts the "clarity" of that image. This is true for detected Z axis features as well. Thus the "image" data of a feature is optimized visually at a Z dimension associated with maximum rate of change of the signal in the Z dimension. This is achieved for each channel by moving the emitter (and receiver) within a quarter wave length in the Z direction. This position is referred to as the "Stand-Off" distance.

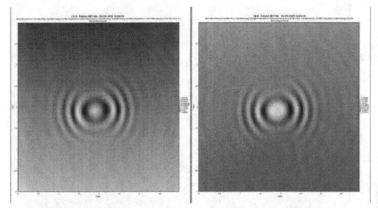

Figure 2.   Comparison of phase difference between channels. Channel A is on left and B on right for a flanged wave guide, with aperture dimensions of 3.96 x 10.67 mm (0.156 x 0.420 inches). Target is a 2.54mm (0.10 inch) conductive sphere located 19 mm below the surface of a glass plate.

EQUIPMENT CONFIGURATION
        The portable version of the Evisive Scan equipment is shown in Figure 3a. This shows the lap-top-computer with the driving electronics and the scan head. The same computer and electronics can be coupled in the laboratory to an XY Positioning Table as shown in Figure 4b. Operating, interface

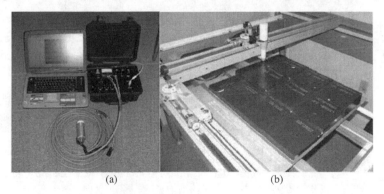

(a)   (b)

Figure 3.   Photographs of portable scanning microwave system
                      a—computer, electronics and head
                      b—X-Y positioning table with head attached

and display software resides on the interface and display computer.

Data are collected via an X-Y raster scan over the surface of the part. The data rate is sufficiently high that mechanical positioning or position feedback for manual positioning is the only limitation in scan speed. The scan data are available in near real time. This scanning technology has been applied in the laboratory, with X-Y planar, X-Y cylindrical and r-θ positioning, and in the field with surface X-Y and multi-degree of freedom positioning devices.

With the exception of the infrared tracking position system, the images presented here were acquired on the X-Y positioning table. Datum spacing in the scan direction was 0.003 inches and raster increment was 0.05 inches unless otherwise stated. Scan speed on the X-Y positioning table reach 3 inches per second and ramp to start and stop. Scan speed with the hand held infrared tracking system varies and may exceed 10 inches per second.

PORTABLE FIELD CONFIGURATION AND WIRELESS HAND-HELD DEVICE:

A number of portable configurations have been applied to field use: The instrument and control system has been interfaced to mechanized pipe scanners and with a multi-axis position system for

Figure 4.   Portable Evisive Scan equipment configured with infrared position tracking

free-form manual positioning, and with a variety of manual position encoders.

A portable system has been interfaced to an infrared camera for correlation in other related studies. The equipment is shown in the field in Figure 4. The probe is manipulated manually, position tracked ad presented in real-time (the position tracking display is shown in Figure 5). The tracking display facilitates control of coverage and scan density. Before and after images are automatically saved with each scan.

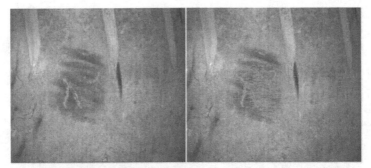

Figure 5.   Scan area image and position tracking image (data points shown in green)

Evisive has incorporated the system electronics into the hand held probe housing and developed a wireless interface of the probe and control computer. This significantly reduces the system size, as well as improving field applicability. The Wireless Hand Held Evisive Scan system is shown in Figure 6.

Figure 6.   Evisive Scan Wireless Handheld System. The Evisive Scan electronics and controls have been incorporated into the Probe Assembly, which communicates by Bluetooth with the User Interface Computer. The camera tracks the infrared beacon providing real time position and data input.

MICROWAVE INTERFERENCE SCAN IMAGES OF ARTIFICIAL LAMINAR DEFECTS

A multi-layer sample panel of ceramic composite armor was examined with several probe configurations. The image data set shown in Figure 7 was obtained with a 10mW 24 GHz transponder and a rectangular wave guide. Channel B of the scan data is shown. The panel has multiple layers of fiber reinforced resin, ceramic tiles of two compositions, two conductive material layers and one elastomeric layer. The panel has six artificial laminar features, arranged at two depths: above and below the elastomeric layer. The scan image in Figure 7 shows that the microwave interferometry system detected all six artificial laminar features. Similar artificial features were detected below (left in the image) and above (right in the image) the elastomeric layer in the part. The difference in geometry of the left and right presentation of these features relates to contours of layers above the deeper features which are on the left in the image. The difference in depth of the features is indicated by the difference in gray scale

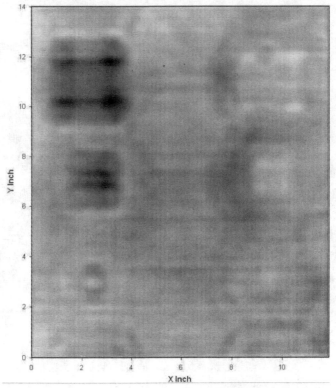

Figure 7.   Scanning microwave image data of specially prepared test sample prepared with artificial delaminations. Six different delaminations in three sizes, located at two depths were inserted and as noted here—these were all detected.

(voltage) which relates to their relative phase positions.

The phase position of laminar features can be adjusted to "focus" the acquired image data at specific depths in the material, or to minimize the effect of specific laminar features. This is particularly beneficial in optimizing the technique for detection of laminar features at a specific layer in the complex material structure.

The small circular indication in the upper right is apparently an unintended laminar feature near the depth of the artificial feature. The difference in depth of the artificial laminar features on the left side (underneath the elastomeric layer) and those on right side (above the elastomeric layer) is apparent in the dark (low negative voltage) and light (high positive voltage) phase difference associated with their depths. The consistent voltage of the three artificial laminar features on the left and right sides similarly indicates their common depth in the material. The circular indication at (9.5, 12.5) is from a laminar feature at a depth very near the artificial feature.

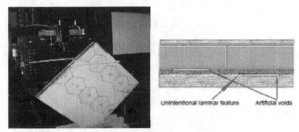

Figure 8.   Photograph of the test panel in the x-ray imaging system, and cartoon of the placement of artificial voids and unintended laminar feature. The upper corner is marked off using a small diameter welding rod to validate the position.

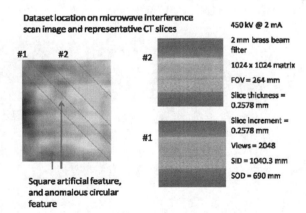

Figure 9.   Correlation between high spatial resolution x-ray computed tomographic images and scanning microwave data. A)-location of x-ray Ct images, b)- x-ray CT section, c)- x-ray CT section. Scan image and CT images of artificial and anomalous features in a sample specimen.

The effect of the interaction with the elastomeric layer above the artificial features on the left is very clear in this image, with the edges of the features occurring at the center of the surrounding patterns.

The artificial feature and analogous feature at (9.5, 12.5) were examined by through transmission x-ray and x-ray computed tomography (CT). The set up for x-ray tomography is shown in Figure 8. A welding rod was placed across a corner to create a temporary position registration.

Examples of the acquired CT images are presented in Figure 9 along with the locations of the CT images relative to the Evisive Scan image data. The complex cross section structure of the specimen is clearly visible in the CT images. While the data was not available at the time of the test, it seems that the thicknesses of the very thin artificial laminar features are below the spatial resolution of the CT image. The anomalous laminar feature is also demonstrated to have a through wall dimension less than the minimum resolution of the CT image (smaller than about 0.25 mm (0.01 inches)).

These experiments demonstrate that the microwave interference scanning method is applicable to complex ceramic armor samples, and that the method seems capable of detecting very thin delaminations at various depths.

DETECTION OF CRACKED ARMOR TILE

The Evisive Scan system has been demonstrated to detect cracked ceramic armor tile in a typical ceramic armor layered configuration. Figure 10 shows a corelation between a through-transmission x-ray image, and twodifferent microwave scan of the same cracked tile. The wide, light gray patterns shown in the microwave scan follow the crack centerlines. The Evisive Scan data has sufficient dynamic range, (greater than 12 bit resoultion), to identify the centerlines and edges of features within the data position precision.

HAND HELD COMPARED TO TABLE X-Y PLOTTER

The wireless system is intended for free-hand, hand-held positioning over the part surface. The data from free-hand scanning was compared to data from the laboratory X-Y Positioning table for the same part. As a demonstration of limitation of performance capability; the free-hand scan pattern was irregular and widely spaced (up to about 12 mm (0.5 inches); scan speed varied up to about 25 mm (10

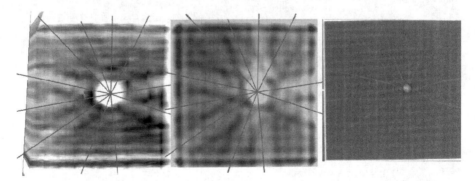

Figure 10.   Evisive Scan Hand Held Infrared position Tracking (left) is compared to Evisive Scan XY positioner (center) and x-ray image (right) of an artificially cracked tile in an armor panel. The detection of the crack pattern is consistent.

inches) per second.

Shown in Figure 10, from left to right are the free hand manually positioned Evisive scan image, the laboratory scan table Evisive scan image and the x-ray image of an artificially cracked tile in an armor panel. The crack pattern in the microwave images is shown by the high (light gray scale) central voltage, and low (dark gray scale) of the interference pattern following the crack. Thus, cracks are indicated by paired diffraction pattern in the Evisive scan images. The central value of the light region corresponds with the white lines of the cracks in the x-ray negative. The diffraction pattern from the tile edges is evident. The prevalence of the horizontal edge pattern in the hand scanned image results from the asymmetrical antenna pattern employed in this test, while the laboratory system was equipped with a probe having an isotropic antenna pattern.

As a demonstration, the hand scanned image shown in Figure 10 has very sparse and irregular data point spacing; up to 12 mm (0.50 inches) in line spacing. The scan lines are not parallel, but have about 2.5 mm ((0.10 inches) data point spacing, which is based on constant sampling rate and variable motion up to about 254 mm / sec (10 inches / sec). This illustrates the practicality of scanning parts in situ under field conditions. Further testing will validate and optimize this capability.

CONCLUSIONS

A portable microwave-based system has been developed and demonstrated on layered ceramic armor to detect cracks and delaminations within ceramic armor systems. Examination requires access from one side only and is effective in applications with metal backing. The capability of the method allows determination of size, depth and orientation of features within the dielectric solid. The laboratory instrument has been successfully coupled to X-Y positioning systems as well as multi-axis scanning systems and free-motion position tracking systems.

The system has been miniaturized and wireless communication incorporated facilitating application in field environments.

Further laboratory testing including destructive analysis of samples will establish scan and data interpretation protocols and qualify the technique for field nondestructive testing applications. The equipment needs further optimization and hardening for field use.

ACKNOWLEDGEMENT

Evisive, Inc. expresses its sincere appreciation to the US Army Small Business Innovative Research Program, and US Army Research Laboratory and US Army Tank Automotive Engineering Research and Development Command.

REFERENCES

[1] J. Salem, D. Zhu, Edited by L. Prokurat Franks, "Advances in Ceramic Armor III", *Ceramic Engineering and Science Proceedings,* Vol. 28, Issue 5, 2007
[2] United States Patent 6,359,446, "Apparatus and Method for Nondestructive Testing of Dielectric Materials", March 19, 2002
[3] United States Patent 6,653,847, "Interferometric Localization of Irregularities", Nov. 25, 2003
[4] International Patent PCT/US2005/026974, "High-Resolution, Nondestructive Imaging of Dielectric Materials", International Filing Date 1 August, 2005
[5] Canadian Patent 2,304,782, "Nondestructive Testing of Dielectric Materials", Mar. 27, 2007
[6] New Zealand Patent 503733, "Nondestructive Testing of Dielectric Materials", PCT/US2005/026974, International Filing Date 1 August, 2005

[7]Australian Patent 746997, "Nondestructive Testing of Dielectric Materials", PCT/US2005/026974, International Filing Date 1 August, 2005

[8]K. Schmidt, J, Little, W. Ellingson, "A Portable Microwave Scanning Technique for Nondestructive Testing of Multilayered Dielectric Materials", *Proceedings of the 32nd International Conference & Exposition on Advanced Ceramics and Composites*, 2008

[9]K. Schmidt, J, Little, W. Ellingson, W. Green, "Optimizing a Portable Microwave Interference Scanning System for Nondestructive Testing of Multi-Layered Dielectric Materials", *Review of Progress in Quantitative NDE*, 2009

CORRECTIVE TECHNIQUES FOR THE ULTRASONIC NONDESTRUCTIVE EVALUATION OF
CERAMIC MATERIALS

S. Bottiglieri and R. A. Haber
Department of Materials Science and Engineering, Rutgers University
Piscataway, NJ, USA

ABSTRACT

When acoustically measuring properties for armor materials, such as alumina or silicon carbide, it is important to incorporate all factors to ensure accurate results. Ultrasonic nondestructive evaluation (NDE) is a widely used technique which makes use of point analysis and property mapping. Material properties usually calculated include attenuation coefficient, ultrasonic velocities, Young's modulus, and other elastic properties. While the overall measurement style is similar between different facilities, several post processing details can be estimated, held constant, or disregarded completely.

Each property which is ultrasonically evaluated needs an input of sample thickness. When mapping elastic property values it is important to have precise thickness measurements at each point. A difference in thickness of only one percent can translate into a variation of material velocity up to 100m/s. Attenuation coefficient, a measure of the loss of ultrasound energy through a sample, is not only affected by the accuracy of knowing sample thickness but is also affected by inherent loss factors. The loss factors which must be incorporated to understand a sample's true attenuation coefficient are reflection and diffraction. By correcting for each of these factors the true values of the material properties can be understood. This paper investigates the variations in acoustic signals due to sample thickness, reflection coefficient, and beam diffraction.

INTRODUCTION

NDE techniques have rapidly become the standard for the qualitative and quantitative characterization of many types of materials. NDE methods are wide spanning and include optic, ultrasonic, electromagnetic, radiographic, thermographic, x-ray and others. The fabrication of material systems such as colloidal suspensions, metals, plastics, and ceramics regularly employ some form of NDE as a means for production process control. Material property determinations and macro-flaw detection are easily made through any of the mentioned techniques.

Concentrations in research focus on how NDE can be used and improved to understand how fabrication methods and variation contribute to the overall microstructural character of a specific material. The development of evaluation techniques also includes understanding what measurement technologies can be used to determine micro- and macro-structural properties. Data processing and analysis methods are constantly under development to keep up with the breadth of information that is capable of being collected using NDE.

Each NDE type has its specific advantages and disadvantages. This paper focuses on ultrasonic NDE in dense ceramic materials and the correction of certain factors which lead to testing difficulty. Ultrasonic NDE is inexpensive, fast, safe, and energy efficient when compared to the other NDE types. It is a robust testing method which allows for spatially locating flaws, material property mapping, and microstructural characterization. Resolution capabilities are determined by the ultrasonic frequency used and the Young's modulus of the tested material. Current technology allows for longitudinal interactions on the scale of tens of microns in dense ceramic materials with frequencies above 100MHz. Error in ultrasonic property measurements is introduced due to sample thickness variation and the acoustic wave phenomena reflection and diffraction. Methods for the improvement in the reduction of measurement error for ultrasonic NDE are further explained.

BACKGROUND

Ultrasonic nondestructive evaluation can make use of several ultrasonic equipment setups. The corrections used for these techniques are for a specific type of lab setup and can be easily modified to accommodate the other types of ultrasonic equipment. The remainder of this paper refers to a pulse-echo, immersion-based ultrasonic scanning system. A pulse-echo system utilizes a single ultrasonic transducer to both transmit and receive the acoustic waves. An immersion-based setup has the sample and transducer face submerged in a fluid medium (in this case, distilled and deionized water) to ensure the most optimal transmission of ultrasound energy into the sample. Other setups may involve the physical contact of transducer to sample using a couplant lubricant or the use of two separate transducers to either transmit or receive. A pulse-echo setup can be considered as the standard for present day ultrasonic flaw detection techniques [1].

A digital oscilloscope is used to display the ultrasonic reflections caused by sample interactions and captured by the transducer. Figure 1 shows an oscilloscope view of ultrasonic pulse-echo reflections and their correspondence to a sample.

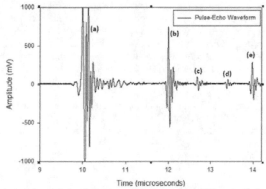

Figure 1. Example of pulse-echo waveform. Reflections of (a) Top surface; (b) Longitudinal peak of 1st bottom surface; (c) Once-through mode-converted reflection from the sample back surface (d) Shear peak of first bottom surface; (e) Longitudinal peak of 2nd bottom surface.

The two types of waves that interact with the ceramic materials looked at are longitudinal and shear. The oscilloscope allows for the measurement of the amplitude and arrival time, or time-of-flight (TOF), for any of the reflected longitudinal and shears peaks which correspond to a specific sample surface. Amplitude and TOF data can then be used to quantify the material acoustic loss and elastic properties, respectively. Common modes of measurement are known as A-scan, B-scan, or C-scans. An A-scan is essentially a point measurement taken directly from the oscilloscope. A-scan data consists of single points of information measureable by the oscilloscope, such as peak amplitude or temporal position. A B-scan is a cross section made up of A-scan data across one axis of a sample. And a C-scan is a two-dimensional downward view consisting of A-scan data over all points of a sample [2, 3].

Elastic properties may be measured by using both the longitudinal and shear TOF's; TOF_L and TOF_S, respectively. This is done by measuring the time between the top surface peak and the first bottom longitudinal peak for LTOF and the time between the top surface peak and the shear peak. Due to the pulse-echo method used, these times are twice the time it would take the ultrasound pulse to travel only once through the material. Equations for the longitudinal velocity, shear velocity, and Young's modulus (E) are shown [4].

$$C_{L,S} = \frac{2t}{TOF_{L,S}} \tag{1}$$

$$E = \frac{\rho \cdot C_L{}^2 \cdot (1-2\nu)(1+\nu)}{(1-\nu)} \tag{2}$$

The factor of 2 in equation 1 is added to account for the pulse-echo configuration and t is the thickness of the sample. In equation 2, ρ is the sample density, and ν is the Poisson ratio obtained from the longitudinal and shear velocities [4]. The terms including the Poisson ratio in equation 2 are to account for the anisotropy typically seen in dense ceramic materials.

Ultrasonic attenuation coefficient is a material property that is integral in understanding the interrelations between microstructure and mechanical properties [5]. It is a quantification of the amount of acoustic energy that is lost as it propagates through the material. It is a measurement that can be made by considering the ratio of the amplitudes of the second bottom surface reflection to the first bottom surface reflection. The ultrasonic amplitude that can is measured for each peak is the square root of intensity. Using this with the Beer-Lambert equation (equation 3), the attenuation coefficient value, α, can be calculated as:

$$\alpha = -\frac{1}{2t}\log\left(\frac{I}{I_0}\right) \tag{3}$$

Attenuation coefficient is typically measured in units of dB/cm, t is sample thickness in cm, I/I_0 is square of the amplitude ratio (A/A_0), and a division by 2 accounts for the pulse-echo method used.

It should be noted that the frequency of the interrogating acoustic beam emitted from the transducers are typically not monochromatic; the transducer output usually cover a distribution of multiple frequencies. Transducer manufacturers characterize the output of each transducer by obtaining an ultrasonic reflection off of a standard material which is considered to be a 'perfect' reflector. This captured reflection is then deconvoluted through the use of a Fast-Fourier transform and represented as a frequency vs. intensity graph. The intensity is given as power spectral density (PSD) which is the square of the real part of the signal amplitude [6]. An example of a frequency profile of a 20MHz-central frequency transducer is shown in figure 2.

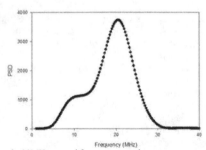

Figure 2. Frequency output range of a 20MHz-central frequency transducer.

A C-scan mapping over an entire sample area of the ultrasonic attenuation coefficient is a weighted average of the attenuation coefficients over all frequencies at each location mapped over an entire sample area. An A-scan, or single point measurement about the sample, of the attenuation coefficient can represented as a function of frequency; this is known as acoustic spectroscopy. A deconvolution of the first two successive bottom surface peaks by an FFT will give two separate graphs of how intense each frequency is at a point. Performing the operation shown in equation 3 at each frequency gives an attenuation coefficient spectrum. Extending this method to all points in a sample to create a C-scan-style map for each frequency is capable of being done through advanced signal and data processing. The end result shows an attenuation coefficient C-scan map at each frequency output by the transducer. This concept is currently being pioneered by the Rutgers University Materials Engineering NDE group. The most common use of this method can be seen in colloidal characterization where it is used to measure particle size distributions in suspensions [7]. Acoustic spectroscopy in metals and ceramics has been investigated to measure grain size distributions, density, and mechanical properties [8, 9].

Ultrasonic NDE is a volumetric test. All of the information that is capable of being produced by acoustic testing requires a thickness input of the sample being examined. A common way of obtaining thickness values is using an average of several caliper measurements. Another way is to assume that the material velocity is constant and solve for thickness, but samples with large anomalous regions within the material will cause a difference in the speed of sound and therefore give an inaccurate thickness measurement. This is detrimental when showing a C-scan image of any material property; slight thickness variations amount to considerable error in actual property contours or values. For example, in alumina, a difference in thickness of only 100μm amounts to a miscalculation of about 7GPa for Young's modulus.

When measuring the attenuation coefficient of a solid material, it is not only thickness that must be accounted for, but also the geometric loss factors: reflection and diffraction. To perform acoustic spectroscopy one must only show the attenuation coefficient that is indicative of how the material causes loss through scattering and absorption. Reflection and diffraction are inherent losses that are caused by the equipment used and acoustic wave phenomena and should not be considered when measuring attenuation coefficients. Equation 4 summarizes what attenuation coefficient, α, must be before making quantifications of a material microstructure [7].

$$\alpha_{Material} = \alpha_{Absorption} + \alpha_{Scattering} = \alpha_{Total} - \alpha_{Reflection} - \alpha_{Diffraction} \qquad (4)$$

THEORY AND DISCUSSION
THICKNESS CORRECTION METHOD

Correction for thickness variation essentially measures time-of-flights around the sample. This method assumes that the speed of sound in water is constant at room temperature and that pure (distilled and deionized) water is used as the propagation medium. The ultrasonic transducer is aligned such that it is perpendicular to the reflector plate controlling the pitch and yaw of the goniometer it is attached to. Once the peak amplitudes are maximized, the goniometer is locked in place and the transducer is unable to move freely during a raster scan. A reflector waveplate that provides a high acoustic impedance mismatch with water is used to ensure that an intense tank peak is seen [10]. The TOF's that are measured are between the face of the transducer and the top surface of the sample (t2), the bottom surface of the sample and the tank containing the sample (t3), and the face of the transducer and the tank (t1). The TOF of the sample (t4), if it were made of water, results from subtracting the total of t2 and t3 from t1. Using the relation of thickness equals velocity multiplied by time and the speed of sound in water; one can obtain an accurate thickness value which avoids the complications of microstructural inhomogeneity. Figure 3 shows a schematic representation of the TOF's that are explained. An entire sample area can be mapped for thickness by using this method and rastering over the sample. Figure 4 shows a C-scan example of this.

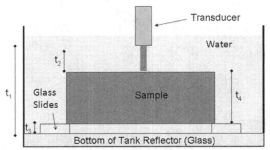

Figure 3. Schematic of test setup and corresponding time-of-flight regions.

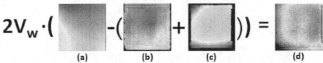

Figure 4. Flow diagram of C-scan thickness mapping. (a) Map of t1; (b) Map of t2; (c) Map of t3; (d) Final thickness map. V_w is the velocity of sound in water and 2 is to account for the pulse-echo method used.

Shown in figures 5 and 6 are examples of C-scan images of an alumina sample. Figure 5 shows the effect of accurate thickness measurements on an attenuation coefficient map. Inaccurate thickness values lead to a noticeable percent difference in overall value for attenuation coefficient. Thickness does not appear to play much of a role in the overall contours of the attenuation coefficient maps. The percent difference map for Young's modulus (of the same sample) show large contour variations as opposed to value differences. The overall values from amplitude based measurements are more strongly affected by thickness variations than map contours. TOF based maps show that the contours are more strongly dependent on thickness rather than the values of the elastic properties.

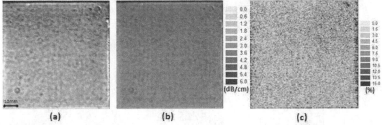

Figure 5. Attenuation coefficient C-scan images of alumina sample. (a) Not thickness corrected; (b) Thickness corrected; (c) Percent difference between (a) and (b).

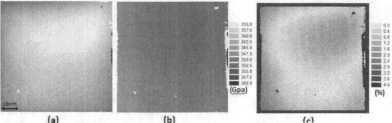

Figure 6. Young's modulus C-scan images of alumina sample. (a) Not thickness corrected; (b) Thickness corrected; (c) Percent difference between (a) and (b).

REFLECTION CORRECTION METHOD

Ultrasonic reflection and transmission occur at boundaries where there is a mismatch between the acoustic impedances of the two material's that form the boundary. The acoustic impedance of a material is characterized as the opposition to displacement of its particles by sound. Mathematically, it is the materials density multiplied by its acoustic velocity. The quantification of acoustic reflection (R) is shown in equation 5, where Z is the acoustic impedance and subscripts represent different materials. The transmission (T) and reflection (R) coefficients are compliments of each other.

$$R = \frac{(Z_2 - Z_1)^2}{(Z_2 + Z_1)^2}$$ (5)

When measuring amplitude-based parameters, it is convenient to use the first two successive bottom surface reflections (henceforth labeled as A_0 and A). This eliminates roughness effects from the top surface and allows for amplification gain to be increased to an appropriate level for better peak resolution [11]. The first reflection peak seen in the oscilloscope image is the reflection off of the top surface of the sample; it represents the ultrasound energy that never penetrated the sample. The subsequent reflections are those which have passed through the bulk of the sample at least twice. Hence, to measure the energy from A_0 and A, considerations must be made to account for the percentage of energy (of the original output) that these peaks actually represent. The measured ratio of the amplitudes of A_0 and A, A/A_0, is actually a percentage of the total energy due to the energy loss at the first surface reflection. An iterative algebraic process (outlined in figure 7), which carefully accounts for the reflection and transmission coefficients at each surface, yields a more accurate form of the equation for measured ultrasonic attenuation coefficient [12].

$$\alpha_{Measured} = -\frac{1}{2t}\left(\log\left(\frac{A}{A_0}\right)^2 + \log(R)^2\right)$$ (7)

The exponents of 2 are to account for intensity being the square of amplitude. This process has been automated to account for reflection while a C-scan is taking place.

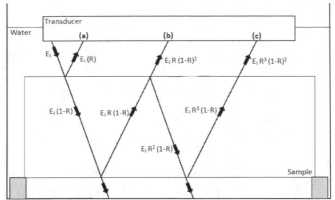

Figure 7. Schematic diagram of acoustic reflection and transmission processes through a solid sample. E_I is initial energy output from transducer. Oscilloscope temporal positions corresponding to (a) Top surface reflection; (b) First bottom surface reflection; (c) Second bottom surface reflection.

Figure 7 demonstrates the processes of ultrasonic reflection and transmission through a dense elastic solid. If one were to consider a situation of an ideal material that causes no loss of the ultrasonic energy and that diffractive losses did not occur, there would still be attenuation caused by reflection phenomena. This attenuation can be described as a quadratic combination of the reflection and transmission coefficients multiplied by the initial amount of energy emitted from the transducer. After the first reflection off of the top surface, the subsequent bottom surface reflections decrease by a factor of R^2 upon each pass. Correction for the attenuation caused by reflection amounts to a linear decrease in attenuation coefficient for both C-scan images and attenuation coefficient spectra by a factor of:

$$\alpha_{Reflection} = -\frac{1}{2t}(\log(R)^2) \qquad (8)$$

To obtain the most accurate values for R about an entire sample area, one should not assume it is constant. The local average modulus, at each point in a sample, can be obtained (from an A-scan) to calculate Z at each point. The acoustic impedance at each point can then be used to distinguish and account for any variation seen in R.

DIFFRACTION CORRECTION METHOD

Diffraction is the phenomena in which sound waves spread out or bend at an interface or discontinuity. The causes of diffraction inherent with a non-contact, pulse-echo ultrasonic NDE system are due to transducer aperture size, path length of beam travel, and the frequencies used [13]. The diffraction error is related to the ratio of the transducer aperture size and the frequency used [14]. Error is increased with smaller apertures and higher frequencies [14]. After the acoustic beam exits the transducer there is a natural Gaussian focusing effect as the beam transitions from the near field to the far field [3]. This becomes amplified by any refraction that occurs at the water-sample surface. By the time the ultrasonic pulse is reflected back, a considerable amount of energy has beam directed away from the transducer aperture. This loss in energy is not due to the material and should not be convoluted with an attenuation coefficient measurement.

The mathematics used in the correction for diffractive losses invoke the usage of spherical Bessel functions and other non-linear transforms. The actual algorithms used for correction are highlighted in the referenced literature [13, 14]. A custom-made program performs the rigorous operations to calculate diffraction losses based on sample thickness, water path, frequency, and transducer aperture radius. When done at every frequency, the overall effect is a subtraction of diffraction losses from the measured attenuation coefficient; shown in figure 8. Figure 8 also demonstrates the frequency dependence of diffraction: as frequency increases, diffractive losses begin to become negligible. Equation 9 shows the diffraction-corrected, modified, version of equation 7. Attenuation due to diffraction is labeled as the ratio of D_1 to D_2, which are complex functions of spherical Bessel operators (shown below).

$$\alpha_{Measured} = -\frac{1}{t}\left(\log\left(\frac{A}{A_0}\right) + \log(R) + \log\frac{D_2}{D_1}\right) \tag{9}$$

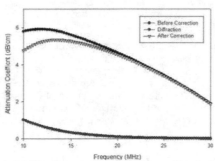

Figure 8. Example of frequency-based diffraction correction on attenuation coefficient values.

To account for the attenuation due to beam spreading one must use a spherical coordinate system and a solution of Bessel's differential equation, the spherical Bessel function of the first kind. It can be represented as an integral of trigonometric functions, but more concise notation will be used here. The Bessel function of the first kind will be shown as $J_m(x)$ in its generalized form. Transforms must be used for each surface that is of interest. The surface peaks that are typically used are the first and second bottom surface reflections, D_1 and D_2, respectively. Shown in equation 10 is the form used to calculate the intensity of the diffracted portion of the beam [13, 14].

$$D_n = 1 - e^{\left(\frac{-2\pi d_n pi}{s}\right)}\left[J_0\left(\frac{2pi}{s}\right) + J_1\left(\frac{2pi}{s}\right)\right] \tag{10}$$

Where $J_{0,1}$ are representative of the first kind of Bessel's functions that account for zeroth and first order perturbation. The subscript 'n' on 'D' is indicative of which surface reflection the transform is manipulating. 's' is defined in equation 11:

$$s = \frac{2\pi \delta_m \cdot C_w + 2 \cdot m \cdot \delta_m \cdot C_m}{f \cdot a^2} \tag{11}$$

The path length of the sample and the water between the transducer face and the top surface of the sample is labeled as t_m and t_w, respectively. The speed of sound in the sample and in water is shown as 'c' with appropriate subscripts. 'f' is frequency, 'a' is transducer aperture radius, and 'n' is either 1 or 2 to correlate with the 1st or 2nd bottom surface reflections.

An attenuation coefficient C-scan map, which is incapable of showing the frequency dependence that exists, can also be corrected for the overall effect of diffraction. As an attenuation coefficient C-scan map represents acoustic loss as the weighted average over the frequency output of a transducer, a weighted average accounting for the effects of diffraction can be calculated. This is done by using the frequency profile of the transducer output and weighting the frequency-dependence of diffraction to the corresponding strength of each emitted frequency. The overall-diffraction correction is then a summation of the weighted frequency-diffraction corrections.

Overall diffraction losses are also dependent on thickness. Custom-made software accounts for this thickness dependency during a rastering C-scan. Figure 9 shows an example of this thickness dependency on overall diffraction corrections in aluminum oxide. The regression line shows what is to be expected: attenuation coefficient is a material property and should not vary with thickness. Figure 10 shows comparative C-scan attenuation coefficient maps of alumina before and after correcting for overall diffraction losses. The effect is essentially a decrease of approximately 0.5dB/cm across the entire sample.

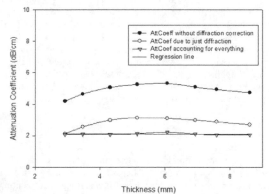

Figure 9. Thickness dependency on overall diffraction correction.

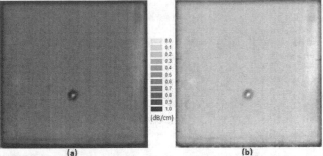

Figure 10. (a) Attenuation C-scan image of an Al$_2$O$_3$ sample without diffraction correction. (b) Attenuation C-scan image of the same Al$_2$O$_3$ sample with diffraction correction.

CONCLUSION

This paper outlines three methods to correct for when ultrasonically measuring the properties of dense ceramic solids. Every material property capable of being measured requires a thickness input. Slight variations in the actual thickness of a sample can lead to inadequate property measurements. Attenuation coefficient values (opposed to map contours) are more strongly affected by thickness. Elastic property maps are affected in the opposite way: it is the contours (rather than values) that are more dependent on thickness variation. Geometric losses, reflection and diffraction, are inherently caused by the equipment, frequency, sample thickness, wave reflections, and path length. The correction for these complications allows for more accurate ultrasonic evaluation. The final governing equation for obtaining the attenuation coefficient of just the material itself is shown as:

$$\alpha_{Material} = -\frac{1}{t}\left(\log\left(\frac{A}{A_0}\right) - \log(R) - \log\frac{D_2}{D_1}\right) \qquad (12)$$

The possibility of ultrasonic nondestructive evaluation to be used as a diagnostic test to spatially locate large flaws has been realized. The use for quantizing material properties in dense ceramic solids has become commonplace is many laboratory settings. Corrections for any type of error that can convolute these measurements have become more important as requirements for specific properties become more stringent. The use of acoustic spectroscopy to understand feature size distributions within elastic solids is becoming an accurate test due to the elimination of geometric loss error.

ACKNOWLEDGEMENTS

The authors would like to thank the NSF IUCRC Ceramic and Composite Materials Center and the Army Research Laboratory Materials Center of Excellence for Lightweight Vehicular Armor.

REFERENCES

[1] Woo, J. "A Short History of the Development of Ultrasound in Obstetrics and Gynecology: Part 1" A Short History of the Development of Ultrasound in Obstetrics and Gynecology. Created 2002. Oxford University. Accessed 2009 < http://www.ob-ultrasound.net/history1.html>.

[2] Brennan, R., "Ultrasonic Nondestructive Evaluation of Armor Ceramics", Ph.D. Thesis, Rutgers University (2007).

[4] D.E. Bray, R.K. Stanley, Nondestructive Evaluation: A Tool in Design, Manufacturing, and Service, CRC Press, Inc., Boca Raton, FL, 1997.

[5] A.E. Brown, "Rationale and Summary of Methods for Determining Ultrasonic Properties of Materials at Lawrence Livermore National Laboratory", Technical Report

[6] Vary, A. and H. Kautz (1987). "Transfer Function Concept for Ultrasonic Characterization of Material Microstructures." Materials Analysis by Ultrasonics – Metals, Ceramics, Composites, Noyes Data Corporation, Park Ridge, New Jersey, pg. 249-289.

[7] Papoulis, A., Signal Analysis. McGraw-Hill, Inc., USA, 1977.

[8] Dukhin, A. and P. Goetz, Ultrasound for Characterizing Colloids – Particle Sizing, Zeta Potential Energy, Elsevier Science B. V., Amsterdam, Netherlands, 2002.

[9] Nicoletti, D. and A. Anderson (1996). "Determination of Grain-Size Distribution From Ultrasonic Attenuation: Transformation and Inversion." Journal of the Acoustical Society of America **101**(2): 686-689.

[10] Serabian, S. (1987). "Ultrasonic Material Property Determinations." Materials Analysis by Ultrasonics – Metals, Ceramics, Composites, Noyes Data Corporation, Park Ridge, New Jersey, pg. 72-78.

[11] Roth, D. (1996). "Commercial Implementation of Ultrasonic Velocity Imaging Methods via Cooperative Agreement Between NASA Lewis Research Center and Sonix, Inc." Nasa Technical Memorandum, 107138.

[12] Lynnworth, L., Ultrasonic Measurements for Process Control. Academic Press, Inc., San Diego, CA, 1989.

[13] Papadakis, E.Physical Acoustics, Principles and Methods, vol. 4, part B, pp. 269-328, Academic Press, 1968.

[14] Xu, W. "Diffraction Correction Methods for Insertion Ultrasound Attenuation Estimation." IEEE Transactions on Biomedical Engineering. Vol. 40, no. 6, 563-570, (1993).

[15] G. Leveque. "Correction of diffraction effects in sound velocity and absorption measurements", Measurements Science and Technology, Vol. 18, 3458-3462, (2007).

QUANTITATIVE EVALUATION OF STRUCTURAL DAMAGE IN LIGHTWEIGHT ARMOR MATERIALS VIA XCT

William H. Green, Kyu C. Cho, Jonathan S. Montgomery, and Herbert Miller
U.S. Army Research Laboratory
Weapons and Materials Research Directorate
ATTN: RDRL-WMM-D
Aberdeen Proving Ground, MD, USA

ABSTRACT

X-ray computed tomography (XCT) has been shown to be an important non-destructive evaluation (NDE) technique for revealing the spatial distribution of ballistically-induced damage in metals, ceramics, and encapsulated ceramic structures. Previous and ongoing work in this area includes assessment of ballistically induced damage in relatively lightweight individual ceramic targets and ceramic armor panels. In this paper the ballistic damage in a novel Mg alloy sample was completely scanned and extensively evaluated using XCT 2-D and 3-D analysis. Features of the damage were correlated with physical processes of damage initiation and growth. The study of this sample is the first known NDE of ballistic damage in this type of lightweight Mg alloy material by XCT or any other NDE method. XCT scans and analyses of damage in the sample will be shown and discussed. This will include virtual 3-D solid visualizations and some quantitative analysis of damage features.

INTRODUCTION

Magnesium (Mg) is being studied for use in lightweight protection systems. Lightweight materials are typically used in armor panel structures in order to decrease weight without losing ballistic performance. Mg is the lightest structural and engineering metal at a density of 1.74 g/cm^3 that is approximately 1/5, 2/5, and 2/3 the weight of iron, titanium, and aluminum, respectively [1, 2]. Mg alloys are being considered as extremely attractive lightweight materials for a wide range of the Army's future applications where weight reduction is a critical requirement because of its low density. Furthermore, magnesium has good vibration damping capacity [3] and low acoustic impedance characteristics [4] that could be of additional benefit to vehicle applications. XCT is an effective and important NDE technique for revealing internal fabrication characteristics and spatial distribution of damage in material specimens. Previous and ongoing work in the area of XCT evaluation includes assessment of ballistically induced damage in individual ceramic targets, ceramic panels, and metal plates. The purposes of XCT evaluation include characterization and understanding of the detectable fabrication structure and/or damage in the complete 3-D space of the specimen or scanned volume, determination of geometric parameters of fabrication and/or damage features for interpretation and useful engineering data, and correlation of physical structure with fabrication methods and damage features and types with the physical processes of damage initiation and growth. In this paper the ballistic damage in a novel Mg alloy sample was completely scanned and extensively evaluated using XCT 2-D and 3-D analysis.

DESCRIPTION OF SPECIMEN AND DIGITAL RADIOGRAPHY RESULTS

The specimen was an approximately 146 mm (5.7") by 186 mm (7.3") rectangular section from a larger impacted test plate with multiple hits, some of which fully penetrated the plate. The specimen included a single complete penetration and the surrounding area as well as undamaged material farther away. Figure 1 shows two photographs of the rear (exit) side of the specimen. The first photograph (1a) shows the relatively large amount of material around the main through hole that was pushed out from the rear of the specimen. The second photograph (1b) includes a ruler to indicate the size of the hole near the rear face, which is approximately 25 mm across. Digital radiographs (DRs) of the specimen were taken through its thickness and width (edge on) using the 420 keV x-ray tube and linear detector array (LDA) setup in centered rotate-only (RO) mode. The x-ray technique (parameters) of the DRs of the specimen were (400 keV, 2.0 mA) and geometries of source-to-object-distance (SOD) = 750.00 mm and source-to-image-distance (SID) = 940.00 mm. Figure 2 shows two through thickness DRs (a and b) and one edge on DR (c) of the specimen. The first two DRs have been processed or "windowed" differently to accentuate one or some features over others. In the first, the penetration hole itself is emphasized and shows a darker area to the lower left of the hole due to missing material in the front (impact) side of the specimen. In the second image, the overall damage and the perimeter of the pushed out region of material in the exit side of the specimen is emphasized. It shows the main hole and the region with impact side missing material as all black. In the edge on (side) image of the specimen with the impact side on the left, the main hole is evident, as well as severe damage in the middle and rear regions of the specimen and the missing material that was pushed out of the exit side. This is also the first image that shows some of the nature of the approximately parallel lateral, or "petal-like", damage mode in the middle and towards the rear of the specimen, which is significantly more visible in the cross-sectional XCT images.

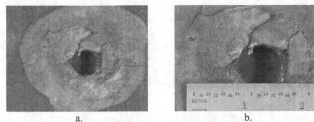

a.   b.

Figure 1. Photographs of rear (exit) side damage in the Mg plate.

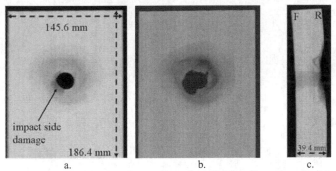

a.   b.   c.

Figure 2. Through thickness (a and b) and edge on (c) digital radiographs of the Mg plate.

XCT SCANNING PROCEDURES
The specimen stood freely on top of a metal plate to raise it up with its exit side facing the x-ray source. Thus, the specimen faces were perpendicular to the horizontal x-ray (collimated) fan beam resulting in through thickness cross-sectional CT images. The bottom (edge) of the specimen was at a vertical position of about 20 mm and the top was at a position of about 206 mm. The middle of the main penetration hole was at a vertical position of approximately 116 mm. The entire volume of the specimen between the vertical positions of 65.000 mm and 173.900 mm was scanned using the 420 keV x-ray tube and LDA set up in offset RO mode. The scans were vertically overlapping with a slice thickness and increment of 0.500 mm and 0.450 mm, respectively, and each slice was reconstructed to a 1024 by 1024 image matrix. The field of reconstruction (FOR) diameter was 195.00 mm. The tube energy and current used were 400 keV and 2.0 mA, respectively, and the focal spot was 0.80 mm. The SOD and SID were 750.00 mm and 940.00 mm, respectively.

QUALITATIVE AND QUANTITATIVE EVALUATION OF SPECIMEN

Computed Tomography Scans
Figure 3 shows a series of CT scans (images) of the specimen with the first at the vertical position of 116.30 mm (3a), which was within 0.20 mm of the position of the center line of the main penetration hole. The scans in Figures 3b and 3c were taken at vertical positions of 126.20 mm and 131.15 mm, respectively, and the scans in Figures 3d and 3e were taken at positions of 106.40 mm and 101.45 mm, respectively. The impact and exit sides of the specimen are at the bottom and top of the images, respectively, and the thickness is 39.4 mm. The missing material towards the front of the specimen adjacent to the main hole is indicated by the shoulder on the right hand side of the cavity wall in Figure 3a. A series of approximately parallel lateral, or "petal-like", cracks on both sides of the penetration cavity are evident. The cracks have the appearance of starting in one direction from the sides of the cavity and then turning or bending back from the original direction towards the rear of the specimen. At some depth into the specimen, the lateral damage mode stops and a relatively large amount of material is pushed out of the exit side. The width of the cavity at the very front and rear of the specimen is 31.78 mm and 90.56 mm, respectively. The narrowest width is 22.27 mm, which is in the middle thickness region. The width of the uppermost cracks towards the rear of the specimen on both the left and right hand side of the penetration cavity is about 1.4 mm. The scans in Figures 3b and 3c are 10.10 mm and 15.05 mm above the main hole center line, respectively. The parallel lateral cracking damage mode is evident in both images. These images do not show a high level of cracking oriented approximately in a through thickness direction. The scans in Figures 3d and 3e are 9.70 mm and 14.65 mm below the center line, respectively. The same lateral damage mode is evident in these images, with a similar relative lack of through thickness cracking.

Three-Dimensional Solid Visualization
The excellent dimensional accuracy and the digital nature of XCT images allow the accurate volume reconstruction of multiple adjacent or overlapping slices. A virtual three-dimensional (3-D) solid image is created by electronically stacking the XCT images, which have thickness over their cross-sections (i.e., voxels), one on top of the other from the bottom to the top of the specimen, or scanned height, to generate its virtual volume. Figure 4 shows a set of four 3-D solid images of the scanned volume with sections virtually removed in 4c and 4d. The method of virtual sectioning, which is essentially only showing a portion of each scan, allows viewing of generated surfaces anywhere in the scanned volume in 3-D space. Figures 4a and 4b show views of the impact and exit side, respectively,

of the entire scanned volume. The very light vertical banding down the middle of the images (top and bottom) is an image artifact from the reconstruction. It is not an indication of a real physical feature in the specimen. The relative shading of lighter and darker gray in the damaged areas is produced by virtual lighting. Physical texture in the surface of the damage in the rear of the specimen is visible in Figure 4b. In Figure 4c, the impact side of the specimen is at the bottom of the image and the sectioned surface is approximately halfway between the top and the bottom of the main penetration hole. In Figure 4d, the impact side of the specimen is on the left side of the image and the sectioned surface is approximately halfway between the left and right sides of the main hole. The relatively lighter surface of the penetration cavity in Figure 4c is due to the angle of the virtual lighting. Both of the surfaces in the sectioned volumes, which are orthogonal to each other, show the parallel lateral cracking damage mode. This is indicative that this damage mode has a significant degree of symmetry about the trajectory of the penetration.

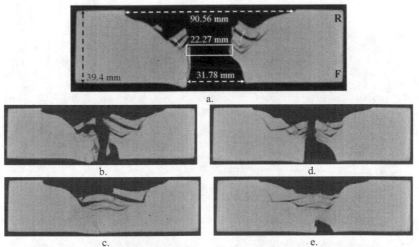

Figure 3. Cross-sectional CT scans (images) of damage in the Mg plate. (a) Vertical position of 116.30 mm. (b) 126.20 mm. (c) 131.15 mm. (d) 106.40 mm. (e) 101.45 mm.

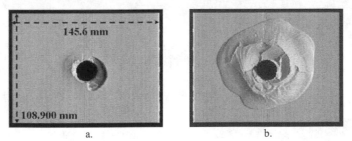

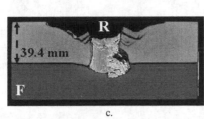

c.   d.

Figure 4. A series of virtual 3-D solid volumes of the damage in the Mg plate. (a) front side. (b) rear side. (c) horizontally sectioned through center of penetration cavity. (d) vertically sectioned through center of penetration cavity.

Three-Dimensional Point Cloud Visualization

A 3-D point cloud is a set of points in space that define geometrical characteristics (i.e., shape, size, location) of a specimen or scanned volume and features within it. Location of the points is determined by appropriate (image) segmentation of the volume or feature(s) of interest. Figure 5 is a point cloud of the overall damage in the specimen and its outside surfaces from a top-down view with the impact side at the bottom of the image. This view clearly shows that there are three distinct regions of different types of damage. The first region of damage towards the front of the specimen with the shoulder on the right is cylindrical and was produced by the initial penetration of the threat. The diameter of the entrance hole without the shoulder included is 21.04 mm with the center at a height of 116.10 mm. The depth and maximum size (parallel to specimen faces) of this region of damage are about 16 mm and 33 mm, respectively. The middle region of damage, which has clear delineation from the damage towards the front, exhibits multiple parallel lateral cracks with "upturned" ends that go towards the rear of the specimen. The depth of this region to the ends of the cracks closest to the exit face of the specimen is about 16 mm. The maximum distance between the ends (tips) of the upturned cracks is about 68 mm. The last region of damage is the relatively large amount of material that was pushed out of the rear of the specimen, which overlaps with the middle region of damage. The physical morphology of the middle and rear damage regions is closely intertwined, making it difficult to determine a precise boundary between the two regions. The maximum size (parallel to specimen faces) of the rear region of damage is about 100 mm.

Figures 6a and 6b are isometric views of the damage point cloud only from exit side and impact side perspectives, respectively, with the boundaries of the scanned volume shown in wireframe mode for reference. The face of the specimen that is away from the view perspective is gridded for ease of interpretation. Figure 7 is an approximate top-down view of the point cloud of the middle region of damage with a small portion of the rear damage to better show the orientation of the lateral cracks relative to the shallow cone of material pushed out the rear of the specimen. The point cloud is tilted backward a few degrees from a top down view in order to separate the upturned ends of the cracks from the surrounding damage as much as possible. Figure 8 is a top-down view of the point cloud of the front region of damage, with some points on the front surface of the specimen to the left and right of the entrance hole. The surface around the entrance hole on the left side is physically raised on the specimen.

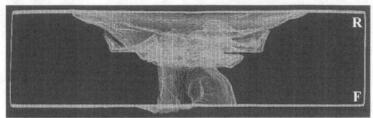

Figure 5. Top-down view of 3-D damage point cloud along with faces and sides of Mg plate (F indicates front and R indicates rear).

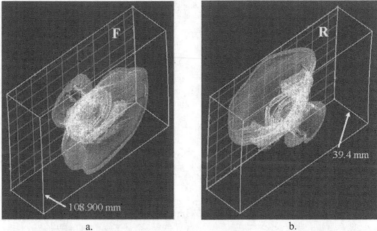

a.   b.

Figure 6. Isometric views of 3-D damage point cloud only with physical boundaries of scanned volume shown as wireframe and face away from the view gridded (F indicates front and R indicates rear).

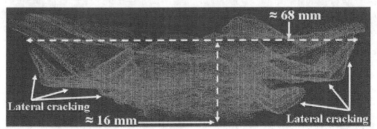

Figure 7. Top-down view of middle section (thickness) of damage point cloud only. The front side of plate is below the bottom of the image.

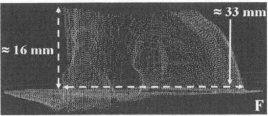

Figure 8. Top-down view of front section (thickness) of damage point cloud only (F indicates front side).

Quantitative Damage Evaluation and Discussion

Image segmentation can also be used to produce binary (black and white) images in which a gray level threshold is applied to separate a feature or features of interest from the material around it. The gray level width of the segmented images is set to two in order to replace each pixel gray level in the original images with a new minimum or maximum gray level (e.g., 0 or 255 for an 8-bit binary image). The pixel data of binary images can be statistically evaluated to determine the level or severity of features, such as the amount of physically detectable penetration damage. This process was done for the XCT scans starting at the bottom of the penetration cavity, including the material pushed out the rear side. The number of pixels in a damaged state in every other scan (slice) ending at the top of the cavity was determined. The matrix size and area of each scan was [1024 x 1024] and (19.50 cm x 19.50 cm), respectively, resulting in an area per unit pixel of 0.036 mm². One approach to representing the damage data is to plot the percent damage relative to the cross-sectional area of the specimen against the vertical distance above and below the center line of the penetration cavity. This does show the trend in the overall damage from the bottom to the top of the cavity. However, in this case the amount of damage is not in terms of an absolute quantity since the cross-sectional area depends in part on the width of the specimen as cut from the larger test plate. A more informative representation of the data is shown in Figure 9a in which the pixel area, 0.036 mm² from an areal pixel density of 2758 cm⁻², was used to convert number of pixels in a damaged state to total damaged area and plot it versus the distance from the center line. Individual segmented binary XCT scans are overlaid on the plot to show the damage at specific locations. The most interesting feature of this plot is the shoulder on the right about 20 mm above the center line. The two binary images shown in this vicinity appear to indicate that less material remained on the left side of the penetration cavity. Figure 9b shows the damaged area plot as well as minimum possible damage from complete penetration and normalized damaged area plots. The minimum damage plot (triangles) is based on the amount of damage that would be present due to a uniform penetration hole through the thickness of the specimen with a diameter equal to the diameter of the base of the threat. Therefore, the minimum damage is zero if the distance above or below the center line is one threat radius or greater. The normalized damaged area plot (squares) is simply the damaged area plot (black diamonds) divided by the minimum damage plot. It shows that the damaged area produced by the penetration at the center line is approximately five times greater than the damage that would exist from a minimum complete penetration hole.

Multiplanar reconstruction (MPR) visualization, also a form of volume reconstruction, used the same set of XCT images as the 3-D solid visualization and was applied to generate individual virtual vertical slices of the specimen from its impact face to its exit face, which were also segmented to binary images. The areal pixel density of views in a MPR image, of which there are four (top, side, front, and oblique), is determined differently than XCT images, since each view is not a single reconstructed image to a set matrix size like a CT image. In this case, the physical area of the specimen and the average number of pixels in that area in the vertical slices (front views) of the MPR images was (14.56 cm x 10.89 cm) and 109,114 pixels, respectively. This resulted in an areal pixel density of 688 cm^{-2} and inversely a pixel area of 0.145 mm^2. Figure 10a is a plot of the damaged area parallel to the faces of the specimen versus depth from its impact face. Individual segmented binary vertical slices are overlaid on the plot to show the damage at specific locations. The plot has two local maxima and three local minima. This behavior of local peaks and valleys is due to the parallel lateral cracking damage mode in the middle of the specimen. The local spatial periodicity of the minima to minima (two) and maxima to maxima (one) segments of the damage is about 5 mm. The plot also shows that the area of the penetration hole necks down to a minimum at a depth of about 14 mm and a steep rise after a depth of about 26 mm, which is indicative of the large amount of material pushed out the back of the specimen. Figure 10b shows the damaged area plot as well as normalized minimum and normalized damaged area plots. Again, the normalized minimum plot (triangles) is based on the amount of damage that would be present due to a uniform penetration hole through the thickness of the specimen with a diameter equal to the diameter of the base of the threat. In this case, the minimum possible damage area as a function of depth is a constant, so this plot is normalized to one. The normalized damaged area plot (squares) is the damaged area plot (black diamonds) divided by the constant minimum damage area. In this way, the normalized damaged area plot gives a factor difference between the actual damaged area and the minimum possible damage with complete penetration (factor = 1). For example, at a depth of about 14 mm the damaged area is about nine times greater than the minimum possible damage area for a uniform through thickness hole. Similarly, at the two depths of the local maxima, about 18 mm and 22.5 mm, the damaged area is about twenty and thirty seven times greater than the minimum, respectively. The through thickness damaged area is at least approximately one order of magnitude or greater than the minimum possible damage area throughout the penetration cavity.

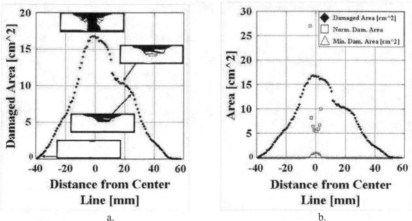

a.   b.

Figure 9. Plots of damage vs. vertical distance from center line of penetration cavity (positive indicates above and negative indicates below center line). (a) damaged area perpendicular to faces. (b) damaged area, minimum damage area, and normalized damaged area on same plot for comparison.

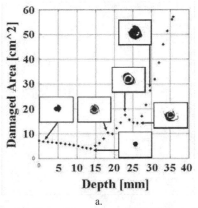
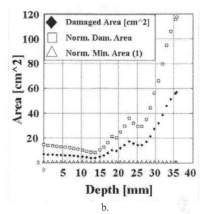

a.   b.

Figure 10. Plots of damage vs. depth (distance from front face). (a) damaged area parallel to faces. (b) damaged area, normalized minimum area, and normalized damaged area on same plot for comparison.

CONCLUSIONS

Ballistic damage in a sectioned specimen from a larger novel Mg plate was scanned and extensively characterized using XCT 2-D cross-sectional (planar) and 3-D volumetric analysis. Damage features including near parallel lateral cracking with "upturned" ends away from the penetration cavity, narrowing and widening cylindrical section of the penetration cavity, asymmetric missing material on one side of the impact face, relatively large area removal of material on the exit side, and three zones of different types of damage and features were captured and discussed. Successive application of XCT 2-D evaluation, volumetric solid visualization and analysis, and volumetric point cloud visualization and analysis provided extensive and important qualitative and quantitative data about damage features. The amount of detectable damage both as a function of the distance from the impact side (depth) and the vertical distance from the center line of the penetration cavity was determined and plotted. The damaged area data was also normalized relative to an appropriate area and plotted. Both sets of data (depth and vertical distance) exhibited features in their plots, including an asymmetric shoulder and local minima and maxima, quantitatively reflecting particular characteristics of the damage. Characteristics of captured damage features provided better understanding of the physical processes of damage initiation and growth. Future work is planned to further analyze the damage features and plot data in conjunction with microstructural observations and evaluation using relevant approaches and specimens from the Mg plate.

REFERENCES

[1]E.F. Emley, *Principles of Magnesium Technology*, Pergamon Press, Oxford, England, (1966).
[2]M.M Avedesian and H. Baker, Ed., *Magnesium and Magnesium Alloys, ASM International*, (1999).
[3]K. Sugimoto, K. Niiya, T. Okamoto, and K. Kishitake, Study of Damping Capacity in Magnesium Alloys, *Trans. JIM*, V. 18, pp. 277-288, (1977).
[4]L.P. Martin, D. Orlikowski, and H. Nguyen, Fabrication and Characterization of Graded Impedance Impactors for Gas Gun Experiments from Tape Cast Metal Powders, *Mat. Sci. Eng. A*, V. 427, pp. 83-91, (2006).

STATIC AND DYNAMIC PROPERTIES OF Mg/CERAMIC MMCs

M. K. Aghajanian, A. L. McCormick, A. L. Marshall, W. M. Waggoner, P. K Karandikar
M Cubed Technologies, Inc.
1 Tralee Industrial Park
Newark, DE 19711

ABSTRACT
    Owing to their attractive properties, particle reinforced, metal matrix composites (MMCs) have shown viability in many applications, including components for thermal management, precision equipment, automotive, aerospace, wear and armor. The vast majority of work to date has examined Al-based MMCs. The present work studies Mg-based MMCs. Mg, due to its low density compared to Al, is more attractive for weight critical applications. Mechanical and physical properties of these materials can be tailored via control of reinforcement chemistry, size, and content; alloy chemistry; and heat treatment. The present study produced Mg MMCs with three different reinforcement chemistries (B_4C, SiC and Al_2O_3). Microstructures, static properties, and dynamic properties (i.e., ballistic resistance) were characterized. Finally, to understand the effect of reduced density, a comparison to results obtained on similar Al-based MMCs was made.

INTRODUCTION
    Discontinuously reinforced, aluminum-based, metal matrix composites (Al MMCs) have received significant attention for various structural applications due to their improved mechanical and wear properties when compared to unreinforced aluminum, their lower density relative to ferrous metals, and their lower cost relative to fiber-reinforced metals and most ceramics. For weight critical applications, such as aerospace, automotive and armor, advanced materials with further reduced density are desired. To this end, the present study examines the fabrication and properties of magnesium-based metal matrix composites (Mg MMCs) for structural applications. The use of magnesium, a very low density structural metal, in combination with low density reinforcement particles allows the production of lightweight, high specific stiffness components. For reference, Table I provides density data for common structural metals and typical ceramic reinforcement phases. Clearly, the combination of a Mg matrix and B_4C reinforcement particles would lead to an attractive low density.

Table I: Density Comparison for Various Matrix and Reinforcement Options

Density Comparison (g/cc)	
Matrix Options [1]	Reinforcement Options [2]
Mg = 1.74	B_4C = 2.52
Al = 2.70	SiC = 3.21
Ti = 4.51	AlN = 3.26
Steel = 7.87	Al_2O_3 = 3.99

    Many methods have been studied for the production of ceramic particle reinforced Al and Mg MMCs, including casting (mixing of molten metal and ceramic powder, followed by a standard foundry practice) [3-5], powder metallurgy (mixing of ceramic and metal powders, followed by solid state hot pressing) [6], pressurized liquid metal infiltration of ceramic particle preforms [7-9], and pressureless liquid metal infiltration of ceramic particle preforms [10, 11].

79

The casting technique is fairly mature and capable of producing large, complex-shaped components, but it is limited to lower particle loadings (typically 30 volume percent or less). The powder metallurgy technique produces high quality composites over a modest range of particle loadings, but is limited with respect to part size and complexity due to hot pressing constraints. By pressurized infiltration of molten metal into ceramic particle preforms, high reinforcement content MMCs with attractive properties can be produced, but size and shape are again limited due to the high processing pressures and complex molds required for manufacture. The last method, pressureless molten metal infiltration of ceramic particle preforms, allows for high particle loadings and the production of complex-shaped components, and thus was used for the present study. This allowed for production of large components without the need for complex molds or high cost capital equipment. The present study examines the fabrication and properties of magnesium-based metal matrix composites (Mg MMCs) for structural applications.

EXPERIMENTAL PROCEDURES

Magnesium MMC composite billets were fabricated to examine the variable of reinforcement chemistry, namely Al_2O_3, SiC and B_4C, on properties. In each case a binary Mg-Al alloy (nominally Mg-11Al – i.e., magnesium casting alloy AM100A) was infiltrated into a 150 mm x 250 mm x 20 mm compact of reinforcement material. The reinforcement content (preform green density) was fixed at nominally 50 volume percent for all three reinforcement chemistries. Constant magnification SEM photomicrographs of the three reinforcement chemistries are provided in Figure 1. All three purchased powders were intended to be identical in particle size with a median particle diameter of nominally 50 microns. However, some differences between the powders are evident, with the Al_2O_3 particles being larger than the SiC and B_4C particles, and the SiC particles being more blocky in shape than the somewhat acicular Al_2O_3 and B_4C particles.

Figure 1: SEM Photomicrographs of Three Particle Chemistries

Infiltration of the matrix metal occurred by coupling the ceramic powder compacts with the Mg-Al alloy in an atmosphere of flowing nitrogen at 700°C (Figure 2). The furnace was then cooled to room temperature with a gradient to allow directional solidification. This controlled the solidification process and yielded composites free of solidification porosity.

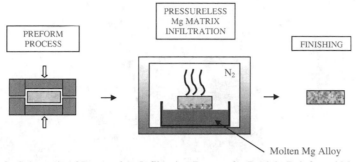

Figure 2: Schematic of Pressureless Infiltration Process for Particle-Reinforced Mg MMCs

TEST PROCEDURES
Physical and mechanical properties were measured with the test methods shown in Table II. Density was measured only once per material on the bulk billet of composite. Young's modulus was measured at three locations on each billet. Flexural strength was measured with five to ten samples using a Sintech universal test frame in conjunction with Test Works materials testing software. Hardness was measured on a ground surface using a minimum of eight indents. Ballistic performance (multi-hit V_{50}) was determined with a minimum of six shots in accordance with military standard. Microstructures were evaluated using a Leica D 2500 M optical microscope and the Clemex Vision PE imaging software. Fracture surfaces were examined using a Jeol JSM-6400 SEM.

Table II: Test Methods

Property	Test Description	Test Procedure
Density	Water Immersion	ASTM B 311
Young's Modulus	Ultrasonic Pulse-Echo	ASTM E 494
Flexural Strength	Four-Point Bend	ASTM C 1161
Hardness	Rockwell B	ASTM E 18
Ballistic Performance	Multi-Hit V_{50}	MIL-STD 662

The flexural strength test method employed during this study is primarily suited for brittle ceramic-based materials. The present Mg MMC composites characterized, however, display some plasticity due to the presence of the ductile Mg alloy matrix phase. Therefore, the results of these tests are not fully quantitative. Nonetheless, they are useful for comparison purposes.

RESULTS AND DISCUSSION

The microstructures of all three Mg MMC samples were similar, showing complete wetting of the ceramic particle compacts, including small interstities between particles, and a two-phase Mg alloy matrix. In the Mg-Al alloy system, the two expected phases are α-Mg and $Mg_{17}Al_{12}$ [12]. An example is provided in Figure 3.

Figure 3: Optical Photomicrographs of Mg/SiC MMC
(α-Mg phase is bright and $Mg_{17}Al_{12}$ phase is gray)

A summary of all property data is provided in Table III, with all testing done with the composites in the as-cast (F) condition. The density results follow the expected trend of decreasing composite density with decreasing density of the reinforcement chemistry (Table I). By rule of mixtures, the calculated densities of the 50% loaded Al_2O_3, SiC and B_4C reinforced Mg-11Al MMCs are 2.92, 2.53 and 2.18 g/cc, respectively. This compares to measured values of 2.82, 2.55 and 2.22, respectively (Table III). The measured and calculated results for the Mg/SiC and Mg/B_4C composites are consistent, suggesting full density (i.e., full infiltration) was obtained. However, the measured value for the Mg/Al_2O_3 MMC is somewhat low compared to the calculated value, suggesting the presence of limited porosity (nominally 3 vol. % by rule of mixtures) or lower particle packing.

The Young's moduli for the three ceramic reinforcement materials are nominally 480, 460 and 400 GPa for B_4C, SiC and Al_2O_3, respectively [2]. The trend of Young's modulus results for the Mg MMCs follows this trend with measured values of 165, 156 and 138 GPa for the Mg/B_4C, Mg/SiC and Mg/Al_2O_3 MMCs, respectively (Table III). In particular, Young's modulus of the Mg/Al_2O_3 composite is significantly lower than those of the other composites, again suggesting the presence of porosity or lower particle packing.

As with the Young's modulus results, the measured hardnesses for the composites follow the rank in hardness of the reinforcement phase, with hardness increasing from Al_2O_3 to SiC to B_4C [2]. Moreover, the hardnesses of the composites (88 to 94 Rockwell B) are well in excess of the unreinforced AM100A alloy matrix, which has a Rockwell B hardness of 53 in the as-cast condition [13].

In all cases, the strengths of the Mg MMCs are high (276 to 456 MPa), suggesting good load transfer (i.e., good bond) between the matrix and reinforcement phases. Moreover the strength data show a significant effect of reinforcement chemistry with the Mg/SiC composite having a strength that is nominally 50% higher than those of the other composites.

Table III: Summary of Property Data

Property	Mg/Al$_2$O$_3$ MMC	Mg/SiC MMC	Mg/B$_4$C MMC
Density (g/cc)	2.82	2.55	2.22
Young's Modulus (GPa)	138	156	165
Flexural Strength (MPa)	276	456	310
Hardness (RB)	88	91	94

Figure 4 shows the specific stiffness (E/ρ) of the Mg MMCs with respect to values for typical metal and ceramic materials. Most metals yield a similarly low specific stiffness of nominally 26 GPa/g/cc. Thus for stiffness and/or weight sensitive applications, there are few metallic options. Ceramics, on the other hand, provide very high specific stiffness, particularly in the case of the lower density carbide and boride materials. However, these materials are brittle and therefore have limited utility in structural applications. Since Mg MMCs have specific stiffness values between ceramics and metals, they are expected to be good candidates for weight critical applications in which both high stiffness and damage tolerance (e.g., ductility) are required. Further increase in specific stiffness is possible by increasing reinforcement content of composites above 50 vol. percent.

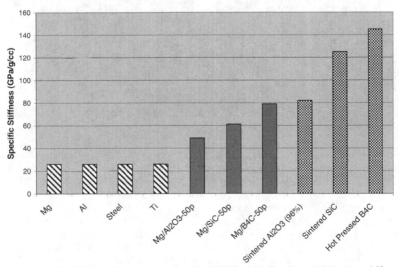

Figure 4: Specific Stiffness Comparison for Mg MMCs with Traditional Metals and Ceramics

Fracture surfaces of all three composites are provided in Figure 5. As expected from the strength data, where a strong bond between phases was suggested, failure is transgranular through the reinforcement particles, with no significant particle pull-out evident. Some indication of ductile failure of the matrix phase can be seen (e.g., knife-edge rupture), particularly adjacent to the particles in the higher strength Mg/SiC and Mg/B$_4$C composites. However, the magnitude of ductile failure is minimal, indicating that the composites are brittle. This finding is consistent with previous work with Mg/SiC composites [5, 9] where low ductility was observed. For the present composites, a significant increase in ductility is likely to be achieved by heat treatment. For instance, a 5 fold increase in elongation of the matrix alloy is attained by changing the condition from as-cast (F) to solutionized and naturally aged (T4) [13].

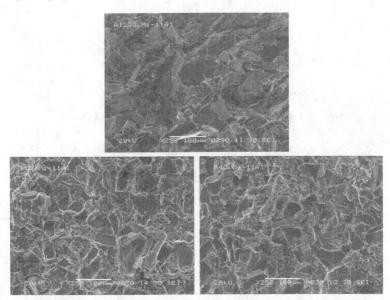

Figure 5: SEM Photomicrographs Showing Fracture Surfaces of Mg MMCs

Al MMCs have been studied as potential armor materials, with attributes including low density, higher hardness than ballistic grade Al alloys, higher toughness than ceramics, ability to erode projectiles, and high specific stiffness [14, 15]. Mg MMCs offer the potential of further performance improvement due to their lower density and high specific stiffness, particularly in the case of Mg/B$_4$C. To assess the potential benefit of Mg MMCs for armor applications, constant areal density, 50% particle reinforced Al/B$_4$C and Mg/B$_4$C MMCs were ballistically tested. Their ballistic performance was evaluated by multi-hit V$_{50}$ testing (50 mm shot spacing) against a small arms armor piercing round using panels measuring up to 400 mm x 400 mm. The results were then normalized to the performance of ballistic grade aluminum (AA 5083). The normalized data (Table IV) demonstrate that both the Al and Mg MMCs provide a performance advantage over unreinforced aluminum. However, despite the lower density of Mg/B$_4$C relative to Al/B$_4$C (2.2 g/cc vs. 2.6 g/cc), the Al/B$_4$C MMC outperformed the Mg/B4C MMC by about

8%. Review of the shot targets showed much higher cracking (i.e., less ductility) in the Mg/B$_4$C sample, indicating that heat treatment of this formulation for ballistic applications is warranted. This brittle failure mode is believed to have hurt performance in the multi-hit V$_{50}$ test.

Similar studies are in process comparing the ballistic performance of unreinforced Al and Mg alloys versus a 7.62 mm armor piercing round [16]. With optimization (processing, composition, microstructure, etc.) it is anticipated that Mg alloys will lead to higher performance due to their lower density. Results to date, however, have shown near identical results for the two different metals on a weight basis [16].

Table IV: Normalized Ballistic Results

Material	Normalized V$_{50}$
5083 Al Alloy	1
Al/B$_4$C-50p	1.4
Mg/B$_4$C-50p	1.3

SUMMARY

Particulate-reinforced, Mg-based, metal matrix composites with various ceramic particulate reinforcement chemistries were fabricated by a pressureless liquid metal infiltration process. Microstructure, static and dynamic properties, and failure mode were evaluated. Key observations can be summarized as follows:

1. Mg-based MMCs with reinforcement phases of Al$_2$O$_3$, SiC and B$_4$C could be successfully fabricated with the pressureless molten metal infiltration process.
2. Due to the very low density of Mg, the resultant Mg MMCs possessed high specific stiffness (e.g., over 3x the specific stiffness of traditional metals for the Mg/B$_4$C MMC).
3. The results indicated the possibility of a strong bond between the Mg matrix and the reinforcement particles, which led to high strength and a transgranular failure mode.
4. In the as-cast (F) condition, the Mg MMCs displayed little ductility.
5. Ballistic performance of the Mg/B$_4$C MMC was good relative to the ballistic grade aluminum baseline, but not as high as that of a similar Al/B$_4$C MMC. Low ductility was theorized for the reduced performance relative to the Al MMC.
6. Of the three composite systems studied, Mg/Al$_2$O$_3$ is most promising for cost driven applications (i.e., addition of low cost Al$_2$O$_3$ particles dilutes the cost of Mg alloy), Mg/SiC is most suited to strength driven applications, and Mg/B$_4$C is best for stiffness and weight critical applications.

REFERENCES

1. *Metals Handbook: Desk Edition* (ASM International, Metals Park, OH, 1985).
2. *Engineered Materials Handbook, Vol. 4, Ceramics and Glasses* (ASM International, Metals Park, OH, 1991).
3. J.T. Burke, C.C. Yang and S.J. Canino, "Processing of Cast Metal Matrix Composites," *AFS Trans.*, **94-179** 585-91 (1994).
4. A. Mortenson, M.N. Gungor, J.A. Cornie and M.C. Flemings, "Alloy Microstructures in Cast Metal Matrix Composites," *J. Metals*, **38** 30-35 (1986).

5. A. Luo, "Processing, Microstructure, and Mechanical Behavior of Cast Magnesium Metal Matrix Composites," Metallurgical and Materials Trans. A, **26** 2445-55 (2007).
6. D.L. McDanels, "Analysis of Stress-Strain, Fracture, and Ductility Behavior of Aluminum Matrix Composites Containing Discontinuous Silicon Carbide Reinforcement," *Metall. Trans. A***16** 1105-15 (1985).
7. J.M. Kaczmar, K. Pietzak and W. Wosiski, "The Production and Application of Metal Matrix Composite Materials," *J. Materials Processing Tech.*, **106** 58-67 (2000).
8. A. Mortenson, "Melt Infiltration of Metal-Matrix Composites," *Comprehensive Composite Materials*, **Vol. 3**, A. Kelly and C. Zweben, Eds. (Elsevier, Amsterdam, 2000), pp. 521-55.
9. V. Kevorkijan, T. Smolar and M. Jelen, "Fabrication of Mg-AZ80/SiC Composite Bars by Pressureless and Pressure-Assisted Infiltration," *Ceramic Bulletin*, 9201-08, June 2003.
10. M.K. Aghajanian, J.T. Burke, D.R.White and A.S. Nagelberg, "A New Infiltration Process for the Fabrication of Metal-Matrix Composites," *SAMPE Q*, **20**(4) 43-47 (1989).
11. M.K. Aghajanian, M.A. Rocazella, J.T. Burke, and S.D. Keck, "The Fabrication of Metal Matrix Composites by a Pressureless Infiltration Technique," *J. Mater. Sci.,* **26** 447-54 *(1991).*
12. *Metals Handbook, Vol. 7, Atlas of Microstructures of Industrial Alloys* (ASM International, Metals Park, OH, 1972).
13. *Magnesium and Magnesium Alloys* (ASM International, Metals Park, OH, 1999).
14. K.T. Leighton, M.K. Aghajanian, R.E. Franz, G.L. Moss and M.A. Rocazella, "Ballistic Performance of Primex-HL Composite Armor," *Proc. Combat Vehicle Survivability Symposium*, **1** 427-40 (1991).
15. E. Chin, "Army Focused Team on Functionally Graded Armor Composites," *Mat. Sci. and Eng. A*, **259** 155-61 (1999).
16. T.L. Jones, R.D. DeLorme, M.S. Burkins and W.A. Gooch, "Ballistic Performance of Magnesium Alloy AZ31B," *Proc. 23rd Int. Symp. on Ballistics*, Tarragona, Spain, 989-95 (2007).

IMPACT STUDY OF AlN-AlON COMPOSITE

Kakoli Das[1], Monamul Haque Dafadar[1], Rajesh Kumar Varma[2] and Sampad Kumar Biswas[1]
[1]Central Glass & Ceramic Research Instt., Kolkata – 700032, INDIA
[2]Terminal Ballistic Research Laboratory, Chandigarh,- 160030 INDIA

ABSTRACT

AlN is well known for its resistance against impact of projectiles. Plasticity of the material under high compressive loading has been well documented in the literature. However, processing of the material is very difficult due to the rapid hydrolysis of the powder. The fabrication of AlN based material utilizing the cheaper commercial powder and the in-situ high temperature reaction has been described in the presentation. The mechanical properties of the material under both quasi-static and impact loading have been reported and correlated with the microstructures. The amount of permanent strain generated in AlN grains after the impact with a 7.62 Ball ammunition has been determined from the X-ray Diffraction line broadening measurement of the debris.

INTRODUCTION

Ceramics are superior to metals in structural applications where density, compressive strength and hardness play important roles. Armors, turbine blades etc are such impact resistant structural components where ceramics are replacing metals due to their possession of superior dynamic strength. Advanced ceramics like B_4C, SiC, AlN, Al_2O_3, AlON (Spinel) are now being tried commercially to offer a better penetration resistant structure. Similarly, binary or ternary composites of these ceramics are now being explored by material scientists for application in these structures primarily due to the combination of properties of the individual components. Ceramic composites being made by novel processing techniques are filling up the gaps in the material – property database for designing applications.

Ceramics undergo plastic deformation when subjected to impact under uniaxial stress or strain conditions. High permanent strain has been observed[1] in confined AlN specimens when recovered after dynamic axial compression test carried out with a modified Split Hopkinson Pressure Bar at a strain rate of $500s^{-1}$. More importantly, the compressive strengths of the unconfined materials increase considerably when the strain rate is increased[1,2]. This means that ceramics could undertake more strains in dynamic conditions than under quasistatic loading conditions. Localisation of shear stresses has been observed from the optical micrographs of the surfaces of post impact Zirconia ceramics[3]. Although inelastic behaviour of ceramics during impact is known for a long time the mechanism of elastic-inelastic transition is still a matter of controversy. Correlation of experimental data obtained from SHPB with those from plate impact experiments revealed[4] a region of brittle – ductile transition in AlN ceramics below the Hugoniot Elastic Limit (HEL) value and a pressure insensitive plastic flow above the HEL value. Recently, it has been shown[5] that plastic flow in the form of dislocation activity could start in high purity and dense Al_2O_3 before the limiting value is reached. HEL value, therefore, indicates only the change in the deformation behaviour from dislocation activity to deformation twinning in this material. The deformation behaviour should also vary with the constituent microstructure of the material and the crystal structure of the phase undergoing deformation.

Microstructure of the debris surface revealed[6] a mixed mode fracture in hot-pressed AlN when subjected to uniaxial compression in the strain rate regime $10^2 - 10^3$ s^{-1}. A micromechanical model based on nucleation, sliding and growth of microcrack within the ceramics has been observed[6] to predict the strain rate dependence of the failure strength of AlN. Post Mortem analysis of the ceramics after the impact with a bullet is very important in finding out the mechanism of plasticity in ceramics.

The present study is, therefore, aimed at finding out the amount of microstrain and stored energy generated in the ceramics after the impact with a 7.62mm NATO ball round launched from an AK-47 rifle and relating the microstructure with the phenomena of microplasticity. The material of the present study is a composite of AlN and AlON(spinel) prepared by the reaction sintering of binary AlN-Al$_2$O$_3$ composition. The impact experiment was carried out in a similar way described in ref.7 except the configuration of the target assembly to look at the behaviour of ceramics under the impact without any energy absorbing backing plate. The microstrain values are calculated for AlN phase.

EXPERIMENTAL PROCEDURE

AlN-AlON composite has been prepared by the reaction sintering of AlN powder coated with hydrated Aluminum Oxide[8,9]. 4 wt% Y$_2$O$_3$ was used as sintering additive and milled with the coated powder using Al$_2$O$_3$ grinding media and isopropanol in a Polypropylene container for 24h. Dried powders were pressed isostatically in a rubber mould at a pressure of 300MPa. Sintering experiments were carried out in a graphite resistance furnace under nitrogen overpressure (0.5MPa) at 2073K for 1h. The heating and cooling rates were maintained at 10K.min^{-1}. Densities of the samples were determined by Archemedes Method. Grinding and polishing of the specimens for metallographic examination were carried out using diamond abrasive slurry. SEM examinations of the polished sections were carried out after coating with carbon for quantitative EDX analysis. Hardness and Fracture Toughness (K$_{IC}$) were determined by using a Vicker's Indenter. Young's Modulus was determined using ultrasound velocity measurement (longitudinal and transverse) and Flexural Strength (4 point) was determined from rectangular bars of dimension 50mm x 5mm x 4mm in Instron testing Machine.

Impact experiments were carried out with an experimental setup shown in Fig.1. Ceramic plates of dimension 50mm X 50mm X 6mm and of the same size and batch were used as targets. The tiles were fabricated by slicing from circular sintered disks. Sample from the remaining portion of the disk was used for the XRD and SEM studies before the impact. 7.62 mm dia NATO Ball round was used as projectile. The launcher was a AK-47 rifle. The mass of the projectile was 9.66g and the velocity was 750 m.s^{-1}. (KE 2.716kJ). The velocity of the projectile was measured precisely from the time of travel between two parallel laser screens placed at a fixed distance (Fig.1). A camera with multiple flash facility was placed at right angles to the line of flight of the projectile which took pictures of the events at a very short time interval. The camera alongwith the flash unit is triggered the moment the projectile passes the second laser screen. The camera lenses are focused on the plane where the event is allowed to happen (line of fire in case of projectile flight). With flashing of each source in sequence light is transmitted through the field lens and focused into the corresponding objective only, to form a shadowgraph of the event on the photographic film. The set up allows exposure time of less than 0.5 microseconds to avoid any blurring. The time delay between the two successive frames can be adjusted from one microsecond to four milliseconds. Thus the system is capable of taking up to one million frames per second. The ceramic plate is placed in between two steel plates which were bolted in four corners and the middle part. A neoprene rubber cloth was placed in between the ceramic and the metal plates primarily to avoid any extra tightening which can produce any tensile force to break the ceramics prematurely. The steel plates had concentric holes (Fig.1) of diameter 25mm at the center for allowing the projectile to impinge on the center of ceramic plate surface directly and were fixed rigidly in the bottom with the help of C clamps. The opposite faces of the ceramic plates were previously ground and polished to a finish of 0.25micron Ra. Tile surfaces were marked to identify the fragments after the impact and know the direction of loading.

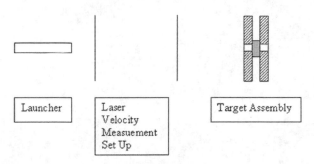

Fig.1 Test Set Up for the Impact Experiment.

RESULTS AND DISCUSSIONS

XRD of the composites sintered at 2073K shows the presence of AlON ($Al_{2.85}O_{3.45}N_{0.55}$) spinel, AlN and YAG ($Y_3Al_5O_{12}$). Absence of α-Al_2O_3 indicated the stability of AlON phase alongwith AlN. AlON formation took place by the reaction,

$$9Al_2O_3 + 5AlN = Al_{23}O_{27}N_5 \tag{1}$$

Part of Al_2O_3 reacted with Y_2O_3 to form YAG by the reaction,

$$5Al_2O_3 + 3Y_2O_3 = 2Y_3Al_5O_{12} \tag{2}$$

Reaction (1) has been observed [10] to take place above 1873K.

The SEM images of polished section of the material are shown in Fig. 2. Primary feature of the micrograph is the absence of any void and the material may be considered fully dense. The densification took place with the help of high temperature liquid phase in the AlN -YAG- Al_2O_3 phase field of AlN-Y_2O_3-Al_2O_3 system[11]. EDX mapping of Oxygen and Yttrium has been presented in Fig.2(c) and (d) respectively to identify the phases distinguished by different electron scattering in Fig. 2(b). Bright white phase contains the maximum amount of Yttrium and Oxygen and, therefore, represents YAG while the dark grey phases do not contain Yttrium or Oxygen indicating AlN. Light grey phases surrounding the dark grey AlN phases contain Oxygen and no Yttrium indicating the presence of AlON. The brightness of the different phases depended on the average atomic number of the elements present in the phases. Core-shell reaction of AlN and Al_2O_3 took place above 1873K to form AlON as evident from the reduction in the sizes of the AlN grains surrounded by AlON grains. Volume content of AlON determined from stereological investigation by a lineal analysis is 25% , while that of AlN is 65% and the balance 10% is represented by YAG.

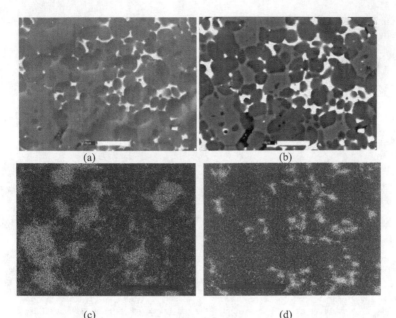

(a) (b)

(c) (d)

Fig. 2. Microstructure of polished sample of reaction sintered AlN-AlON composite (a) secondary electron image; (b) backscattered electron image; (c) Mapping of Oxygen and (d) Mapping of Yttrium. Bars- 10 µm

Mechanical properties of the material are presented in Table I.

Table I Mechanical Properties of the dense AlN-AlON Composites

Density	3310 kg.m^{-3}
Flexural Strength (4point)	164 MPa
Vickers Hardness	11.23GPa (9.8N load)
Fracture Toughness (K_{IC})	2.5 MPam$^{1/2}$
Young's Modulus	302 GPa

Impact events of 7.62mm NATO steel ball on target AlN-AlON ceramic plate are shown in Fig. 3. The events were recorded with an optical sytem as described in the earlier section and clearly show the fragmentation of the impactor (white flashes) , spalling (backward ejection of ceramic rubbles) and communition of the ceramics resulting in production of ceramic rubbles which move at very high speeds in the direction of the launching. Marked portion of the sample recovered from the sandwitch frame after the impact has been subsequently analysed with XRD. Fracture surface of the same portion generated by the impact was analysed by SEM.

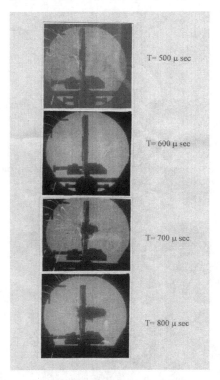

Fig. 3. Impact events recorded by a high speed camera and multiple flash photography techniques. The time was recorded from the moment the projectile passes the second laser beam in Fig.1

Considerable line broadening as well as peak shifts have been observed in the XRD patterns of samples after the impact as presented in Fig.4. Broadening could be found in all the reflection planes. The line broadening is the cumulative effect of instrument, crystallite size and microstrain and can be represnted by the following equation,

$$B_{exp} = B_{inst} + B_{cryst} + B_{microstrain} \qquad (3)$$

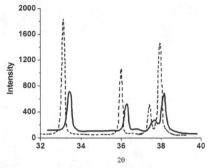

Fig.4: XRD profiles of AIN samples before (dotted line) and after (filled line) the impact.

Since the average grain size determined from the polished microstructure (Fig.2) is much above 1000nm and the earlier observation[18] showing no reduction of particle size in AIN compact after shock treatment the experimental broadening could be represented as

$$B_{exp} = B_{inst} + B_{microstrain} \tag{4}$$

indicating the negligible effect of crystallite size reduction after the impact. Correction for the instrument broadening has been made by standard procedures involving the determinations of the peak widths of sintered and annealed AIN having grain sizes greater than 2 microns under the same experimental set up. B_{corr} becomes the FWHM value due only to the microstrain and could be represented by [12],

$$B_{corr} = B_{microstrain} = 4 \, \varepsilon \tan \theta \tag{5}$$

where 'ε' is the microstrain corresponding to the Bragg angle 'θ'. $B_{microstrain}$ has been determined by PC-APD software after determining the exact peak position. In the present study, $B_{microstrain}$ values have been calculated from the differences of the FWHM (Full Width Half Maximum) values obtained from the samples after and before the impact and are presented in Table II alongwith the microstrains(ε) calculated from eqn.(5).

Table II FWHM and Strains (ε)for different XRD peaks of AIN in the composite observed before and after the impact.

Before Impact			After Impact			B	ε
Hkl	2θ	FWHM	hkl	2θ	FWHM		
100	33.148	0.191	100	33.353	0.405	0.214	0.003116999
002	35.988	0.189	002	36.238	0.406	0.217	0.002893614
101	37.863	0.199	101	38.328	0.426	0.227	0.002850019
102	49.742	0.236	102	50.128	0.57	0.334	0.003116204
110	59.278	0.269	110	59.673	0.674	0.405	0.003081058
103	65.973	0.307	103	66.323	0.692	0.385	0.002570894
112	71.376	0.324	112	71.588	0.785	0.461	0.002789618
201	72.551	0.291	201	72.843	0.802	0.511	0.00302186

Both peak shift and broadening indicate the existence of uniform and non-uniform strains in the impacted ceramics. Average strain value obtained from Table 1 is 2.93 X 10^{-3} in the post impact sample. Energy stored in the lattice of AlN has been calculated from the relation[13],

$$V = 2.81 \, E \, \varepsilon^2 \tag{8}$$

Substituting the value of 'ε', density (3240 Kg m^{-3} for AlN) and E (310GPa for AlN) and from the conversion of 1 MPa = 1001 kJ.m^{-3} the value of stored energy has been found to be 2.302 kJ.Kg^{-1} . This value is higher than that obtained by shock treated powders of AlN^{14}. High value of stored energy in the material is generated due to the inelastic deformation caused by the propagation of shock wave on impact of the projectile.

SEM images of fracture surfaces generated on quasistatic loading (4-pt. flexural) before the impact and on dynamic loading after the impact have been presented in Fig. 5(a), (b) and (c) respectively. There is a difference in the fracture surfaces of the sample. The surface is rough and substantial intergranular fracture are visible in Fig.5(a) when the fracture took place under qasistatic load. Fracture surface becomes planar and mostly transgranular having cleavage steps originating from the microplasticity of the grains could be observed when the load is dynamic. Coalescence of cracks have been produced under dynamic loading [Fig. 5(b)]. A few grains are seen to have extensive microfracture on impact marked by M in Fig.5(c). Evidence of similar features of microplasticity of grains had been observed by earlier workers[5,15] on examination of fracture surfaces of post shock Al_2O_3 by SEM. Cleavage steps have been proposed[5] to be produced in Al_2O_3 ceramics from the deformation twinnings generated in the material on application of pressures above the HEL. Further investigations in the mechanism of cleavage fracture in AlN ceramics under dynamic loading conditions are needed to understand the material model.

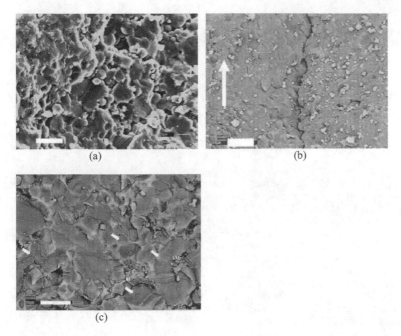

(a)

(b)

(c)

Fig.5 SEM image of fracture surface of AlN-AlON composite (a) before the impact and (b) and (c) after the impact. Arrow in Fig.5(b) indicates the loading direction. White powder like debris of the fragmented metallic projectile could be visible in Fig.5(b) and (c) while a few microfracture (M) and cleavage steps (arrows) in most of the grains are visible in Fig.5(c).

CONCLUSION

Impact analysis of ceramic composite containing 65vol% AlN and 25vol% AlON shows that the material failed inelastically showing transgranular cleavage steps on the application of dynamic load with a 7.62mm dia steel ball (2.71kJ) and AK-47 Rifle whereas brittle fracture showing intergranular features takes place during application of quasistatic four point flexural load. XRD analysis of the material before and after the impact shows the generation of considerable microstrain on impact which contributes to the creation of sufficient stored energy extracted from the projectile kinetic energy.

ACKNOWLEDGEMENT

The authors wish to acknowledge the financial support obtained for this work in the form of Grant-in aid project of DTSR (DRDO).

REFERENCES

1. W.Chen and G. Ravichandran, Static and Dynamic Compressive Behaviour of Aluminum Nitride under Moderate Confinement, *J.Am. Ceram.Soc.*,**79**, 579-84 (1996).
2. J. Lankford, C.E.Anderson Jr, A.J.Nagy, J.D.Walker, A.E.Nicholls and R.A.Page, Inelastic Response of Confined Aluminium Oxide under Dynamic Loading Conditions, *J.Mat. Sci.*, **33**, 1619-1625 (1998),.
3. G.Subhash and S.Nemat-Nasser, Dynamic Stress-Induced Transformation and Texture Formation in Uniaxial Compression of Zirconia Ceramics, *J.Am.Ceram.Soc.* **76**, 153-165 (1993).
4. W.Chen and G.Ravichandran, Failure Mode Transition in Ceramics under Dynamic Multiaxial Compression, *Int.J.Fracture*, **101**, 141-159 (2000).
5. M.W.Chen, J.W.McCauley, D.P.Dandekar and N.K.Bourne, " Dynamic Plasticity and Failure of High Purity Alumina under Shock Loading" *Nat.Mat.* **5,** 614-617 (2006).
6. G.Subhash and G.Ravichandran, Mechanical Behaviour of a Hot Pressed Aluminum Nitride under Uniaxial Compression, *J. Mat.Sc.,* **33** 1933-1939 (1998).
7. J.E.Field, Q.Sun and D.Townsend, "Ballistic Impact of Ceramics", pp 387- 394 in Mechanical Properties of Materials at High Rates of Strain 1989, Proc. 4th International Conf. on the Mechanical Properties Materials at High Rates of Strain, Oxford, UK 19-22March, 1989 , Edited by J.Harding, Instt. Of Physics , Bristol and NY, 1989.
8. K.Das, H.S.Maiti and S.K.Biswas, Indian Patent No. IN200100307-I1, 2008.
9. A.Negi, K.Das and S.K.Biswas, unpublished work
10. H.X.Willems,M.M.R.M.Hendrix,R.Metselaar and G.de With, Thermodynamics of AlON I: Stability at Lower Temperatures, *J.Eur.Ceram.Soc.*, **10** 327-337 (1992).
11. M.Medraj, R.Hammond,W.T.Thompson and R.A.L.Drew, High Temp. Neutron Diffraction of the AlN-Al$_2$O$_3$-Y$_2$O$_3$ system., *J.Am.Ceram.Soc.*, **86**, 717-26 (2003).
12. H.P.Klug and L.E.Alexander in X Ray Diffraction Procedure For Polycrystaline and Amorphous Materials , 2nd. Edition, John Wiley & Sons, NY , 1974, USA.
13. E.A.Faulkner, Calculation of Stored Energy From Broadening of X-ray diffraction Lines, *Phil.Mag.*,**5**, 519-521 (1960).
14. Y. Wu, W G Miao, H P Zhou, W. Han, H. L. Xue, Characterization of shock waves treated aluminium nitride powder and its potential for low temperature sintering, *Mat.Chem.Phys.*, **62**, 91-94 (2000).
15. F.Longy and J.Cagnoux, Plasticity and Microcracking in Shock-Loaded Alumina, *J.Am.Ceram.Soc.*, **72**, 971-979 (1989).

BALLISTIC EVALUATION AND DAMAGE CHARACTERIZATION OF METAL-CERAMIC INTERPENETRATING COMPOSITES FOR LIGHT ARMOR APPLICATIONS

Hong Chang, Jon Binner and Rebecca Higginson
Department of Materials, Loughborough University.
Loughborough, Leicestershire. LE11 3TU, UK.

ABSTRACT

3-3 metal-ceramic interpenetrating composites (IPCs) consisting of 3-dimensionally interpenetrating matrices of two different phases are novel materials with potentially superior multifunctional properties compared with traditional metal matrix composite. The aim of the work underpinning this paper was the understanding of the ballistic performance of the IPCs through dynamic property testing and the subsequent damage assessment. Al(Mg)/Al_2O_3 IPCs having sizes up to 70 mm diameter were manufactured using a pressureless infiltration technique; the samples were then tested using both split Hopkinson's pressure bar (SHPB) and depth of penetration (DoP) methods. Damage in the whole system was assessed via a series of techniques including optical, scanning electron microscopy and transmission electron microscopy, with some samples prepared by dual beam focused ion beam (FIB) techniques.

Though the IPCs contained rigid ceramic struts, they deformed plastically with only localised fracture in the ceramic phase following the SHPB testing. In terms of the effects of the Al_2O_3 content on the ballistic performance of the IPCs, though the IPCs with a higher Al_2O_3 content broke into more fragments, they were more effective in terms of the DoP. Whilst the IPCs were not suitable for resisting high velocity, armour piercing rounds on their own, when bonded to a 4 mm thick, dense Al_2O_3 front face, the component defeated the armour piercing (AP) round and survived fundamentally intact with only minor spall damage in the Al_2O_3 facing.

INTRODUCTION

For an armour tile to be effective, it needs both high penetration resistance and the capability of withstanding more than a single impact, i.e. multi-hit potential. Whilst ceramics such as Al_2O_3, SiC and TiB_2 are attractive materials for ballistic applications in terms of their abrasion resistance, which can blunt the incoming projectiles and absorb the energy, hence defeating the threat, they have poor multi-hit potential, shattering after as little as one impact and needing to be replaced[1]. In addition, most armour systems are made up of composite layers of a number of materials to obtain the maximum protection for the minimum mass. However, when there is a ceramic front face, acoustic impedance mismatches at the resulting interfaces can be a cause of significant problems since the stress waves from the ballistic event are reflected back inside the ceramic as tensile waves, causing its rapid destruction[2].

Metal matrix composites (MMCs) have been shown to display a number of useful properties for a wide range of different applications[3-5]. As a result, they are being increasingly used in applications such as aerospace and defence components[6-7]. Amongst the MMCs, 3-3 interpenetrating composites (IPCs) consisting of 3-dimensionally interpenetrating matrices of metallic and ceramic phases are interesting materials with potentially superior properties compared with traditional dispersed phase composites[8]. By careful control of the thermo-atmospheric cycle and the use of precursor coatings, a range of molten aluminium alloys can be infiltrated without the requirement for pressure into a number of ceramic foam compositions, including alumina, mullite and silicon carbide[9-11].

97

The objective of the present work was to manufacture IPCs using the pressureless infiltration technique and then to evaluate their ballistic properties using both SHPB and DoP approaches. The effect of the density of the precursor Al_2O_3 foams on the subsequent ballistic performance of the composites was studied and the resulting damage to the system was thoroughly assessed.

EXPERIMENTAL

Processing: The precursor Al_2O_3 foams were supplied by Dyson Thermal Technologies, Sheffield, UK. The foams, made by gel casting an aqueous suspension, measured 70 mm in diameter by 10 mm thick and had densities of 15 – 30 % of theoretical and cell sizes in the range of 50 – 150 µm. The Al_2O_3 foam samples were placed on top of similarly shaped and sized discs of an Al-8 wt.% Mg alloy, selected on the basis of previous research[12], and the metal-ceramic couples were then placed in alumina boats and heated at 20°C min[-1] in flowing Ar in a tube furnace; once the temperature reached 915°C the gas was switched to pure N_2 and a holding time of 30 mins was sufficient for complete infiltration. One additional series of samples was made in which individual 4 mm thick, slip cast, dense alumina discs were attached to Al-Mg / 25% Al_2O_3 IPCs (8 mm thick) *in situ* during the infiltration process by providing excess metal which formed an interfacial layer between the ceramic front face and IPC backing.

Ballistic Testing: Initial evaluation of the high strain rate characteristics of the composites was carried out using the SHPB technique on samples measuring 9 mm in diameter and 4.5 mm in thickness. The DoP ballistic evaluation of the composites was performed by Permali (Gloucester) Ltd, Gloucester, UK, using 7.62 mm, steel tipped, armour piercing (AP) rounds at a velocity of 700 ± 20 ms[-1]. In the latter, the composites were glued onto an aluminium backing with a thickness of ~50 mm; the residual energy of the bullet after passing through the target composite was indicated quantitatively by the depth of penetration of the round into the backing. This was ascertained by cutting the backing aluminium in half to reveal the DoP.

Microstructure Characterisation & Damage Assessment: For polarized light microscopy, fragments from the IPCs were ground and polished metallographically using diamond paste and then anodized using fluroboric acid. For scanning electron microscopy (SEM) observation (1530VP FEG SEM, LEO Elektronenskopie GmbH, Oberkochen, Germany), the samples were given a final polish using 0.02 µm colloidal silica prior to the observation. TEM foils (examined using a 2000FX, Jeol, Tokyo, Japan) were specifically prepared from the metal-ceramic interface using a Dual Beam Focused Ion Beam (Nanolab 600, FEI Europe, Eindhoven, The Netherlands) from both SHPB and DoP tested composites; the latter samples were produced from near the impact site.

RESULTS

Ballistic Testing: Representative SHPB results are shown in Fig. 1. In Fig. 1(a), the Al-Mg alloy, with no ceramic foam present, showed a continuous increase in the stress with strain as expected as a result of strain / work hardening. Although the IPCs contained a continuous Al_2O_3 network throughout their structure, they yielded at ~2% strain then displayed plastic deformation behaviour more typical of a metallic material. With an increase in ceramic content in the IPC, the maximum true stress observed increased, the Al-Mg / 30% Al_2O_3 IPC had the highest value of 600 MPa, but the degree of strain decreased. There was no obvious difference in the true stress-strain curves of the Al-Mg / 15% Al_2O_3 IPCs having average cell sizes of 50-100 μm and those of 100-150 μm, implying that cell size had a marginal effect.

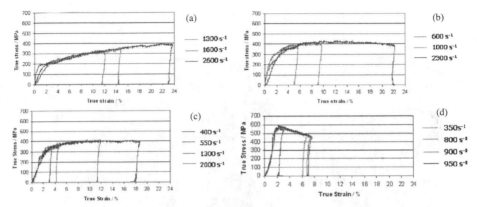

Fig. 1 True stress-strain curves of (a) Al-Mg alloy; (b) Al-Mg/15% Al_2O_3, 50-100 μm; (c) Al-Mg/15% Al_2O_3, 100-150 μm; (d) Al-Mg/30% Al_2O_3, 50-100 μm.

The DoPs of the IPCs after ballistic tests are shown in table 1. The results show that the IPCs on their own are insufficient to stop a high velocity round and the DoP values change little. However, the results with respect to deflection for the IPCs produced from 30% dense Al_2O_3 foam were considered encouraging;once the round hit the composite, the hard, continuous Al_2O_3 network deflected the round and hence contributed to the resistance to penetration.

The results, however, were greatly surpassed by the ballistic performance of the Al-Mg / 25% Al_2O_3 IPC *in situ* bonded with a 4 mm thick, slip cast, dense Al_2O_3 disc, the ceramic – faced IPCs. These samples effectively defeated the AP round with zero penetration into the Al alloy backing, though the samples themselves generally broke into several pieces. One sample, however, survived fundamentally intact with only the Al_2O_3 dense facing cracking perpendicular to the impact and the front of it spalling off, Fig. 2(a). A typical 3D MicroCT micrograph is shown in Fig. 2(b). It is clear that no damage had been caused to the IPC layer beneath the ceramic front face for this sample. The ballistic properties of ceramic-faced armours have been widely studied; the presence of a functionally gradient layer between the ceramic front face and metal back face can make the acoustic impedance change less abrupt resulting in less damage from reflected tensile forces[13]. In this research, the attachment of the IPC to the dense Al_2O_3 layer may have reduced the acoustic impedance mismatch, resulting in the superior ballistic performance of the samples.

Table 1 Results of DoP ballistic tests obtained on the IPC samples.

IPC composition	Avg cell size / μm	Avg DoP* / mm	Deflection
Al-Mg / 15% Al_2O_3	50 – 100	24.7	0-7°
Al-Mg / 15% Al_2O_3	100 – 150	25.3	0-7°
Al-Mg / 30% Al_2O_3	50 – 100	23.5	5-17°

*Depth into aluminium backing alloy.

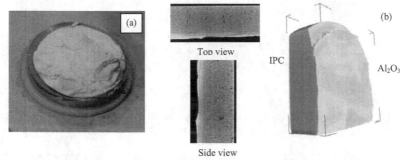

Fig. 2 Micrographs of (a) a ceramic-faced configuration remained mostly intact; (b) a 3D MicroCT micrograph.

MICROSTRUCTURE CHARACTERISATION & DAMAGE ASSESSMENT

Typical optical images and SEM micrographs of the Al-Mg/15%Al_2O_3 IPCs tested under various strain rates are shown in Fig. 3. Though the IPCs contained rigid ceramic struts, the samples deformed plastically with only localised fracture in the ceramic phase. The cracks appear parallel to the compression direction, which indicates that they were formed as a result of tensile stresses. At the higher strain rate of 2300 s^{-1} (Figs. 3(c, d), the compression of the Al alloy is evident; the foam struts broke locally and penetrated into the softer Al, preventing the shattering and fragmentation of the whole sample. The Al-Mg/30%Al_2O_3 IPCs show few microcracks at the strain rate of 350 s^{-1}, Fig. 4(a), though many more may be observed after exposure to the higher strain rate of 900 s^{-1} with localised 'collapse' of the ceramic struts, Fig. 4(b). The IPCs having other foam densities and cell sizes exhibited the similar compression behaviour with only localized cracks formed in the Al_2O_3 struts.

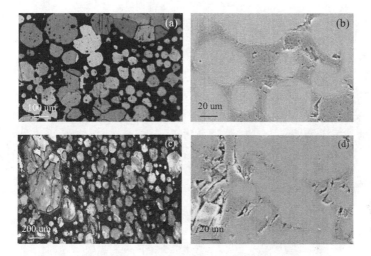

Fig. 3 Optical and SEM micrographs of the Al-Mg/15%Al$_2$O$_3$, 50-100 μm after SHPB testing at (a, b) 600 s^{-1}; (c, d) 2300 s^{-1}.

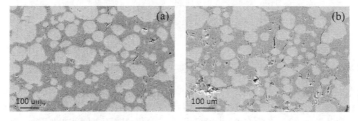

Fig. 4 SEM micrographs of the Al-Mg/30%Al$_2$O$_3$, 50-100 um after SHPB testing at (a) 350 s^{-1} (b) 900 s^{-1}.

Micrographs of the IPCs after DoP testing are shown in Fig. 5. In Fig. 5(a), an IPC made from 15% dense Al$_2$O$_3$ reveals that the original spherical cells were deformed to ellipsoids near the point of impact in a similar manner to the SHPB test (the arrow indicates the general direction of the transmission wave). In contrast, the composites produced from 30% dense Al$_2$O$_3$ showed barely any deformation of the spherical cells, Fig. 5(b), and the metal may be observed to bridge the crack in the image; this latter behaviour must have contributed to the structural integrity and performance of the IPC. Typical SEM cross-section micrographs of a fragment from the ceramic-faced IPCs are shown in Fig. 5(c), where no ceramic - IPC interfacial debonding was observed (indicating the effectiveness of the metal layer), though a few fine cracks in the ceramic front layer may be evident as a result of reflections of longitudinal stress waves[14]. Compared with Figs. 5 (a, b), no obvious deformation of the original spherical cells into ellipsoids, or cracks in the ceramic struts was seen.

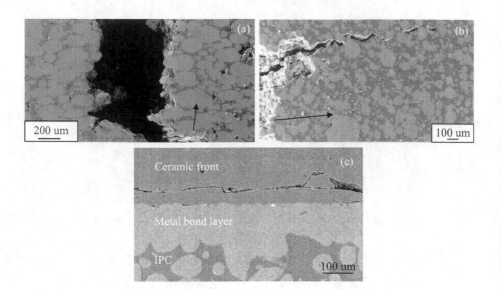

Fig. 5 SEM micrographs of the IPCs after DoP: (a) Al-Mg/15%Al$_2$O$_3$; (b) Al-Mg/30%Al$_2$O$_3$; (c) the ceramic – faced IPCs.

An image of a TEM sample prepared from the Dual Beam FIB at the metal-ceramic interface is shown in Fig. 6(a); from the figure, it can be seen that the interface remained intact without any porosity or interfacial debonding. As with previous research, tremendous numbers of dislocations have been formed in the metal alloy in both SHPB and DoP samples, which is as expected[15]; from Fig. 6(b), a large number of dislocations were again observed in the metallic phase along the Al$_2$O$_3$ grain boundaries, whilst few dislocations were observed in the Al$_2$O$_3$ grains. The formation of the dislocations in the metallic phase along the Al$_2$O$_3$ grain boundaries improved the plastic deformation capability of the IPCs, resulting in the better ballistic properties.

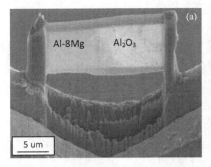

Fig. 6 TEM micrographs of the IPCs after ballistic testing: (a) a TEM sample; (b) the ceramic struts.

CONCLUSIONS

Al-Mg / Al_2O_3 interpenetrating composites have been produced by the pressureless infiltration of Al_2O_3 foams with densities in the range 15 – 30% of theoretical and cell diameters on the order of 50 – 150 μm. The ballistic properties of the IPCs have been assessed using both SHPB and DoP techniques and the resulting damage in the samples evaluated by a range of microscopy techniques. The results have shown that, on their own, the IPCs are not suitable for resisting high velocity, armour piercing rounds, however, when bonded to a 4 mm thick, dense Al_2O_3 front face, the component resulted zero penetration into the Al backing and more importantly, the sample remained mostly intact. Interestingly, the dense ceramic front face did not separate from the IPC even when hit by a steel tipped armour piercing round at 700 ms^{-1}. It is believed that the presence of the IPC layer, between the ceramic front face and the Al backing, reduced the acoustic impedance mismatch.

REFERENCES

[1]K.S. Kumar and M.S. DiPietro, Ballistic penetration response of intermetallic matrix composites, Scrip. *Metall. Mater.*, **32,** 793-8 (1995).
[2]A. Tasdemirci and I.W. Hall, The effects of plastic deformation on stress wave propagation in multi-layer materials, *Inter. J. Imp. Eng.*, **34,** 1797–1813 (2007).
[3]S.C. Tjong and Z.Y. Ma, Microstructural and mechanical characteristics of in situ metal matrix composites, *Mater. Sci. Eng. R.*, **29** 49-113 (2000).
[4]R.B. Bhagat, M.F. Amateau, M.B. House, K.C. Meinert and P. Nisson, Elevated temperature strength, aging response and creep of aluminium matrix composites, *J. Compo. Mater.*, **26** 1578-93 (1992).
[5]M. Manoharan and J.J. Lewandowski, Crack initiation and growth toughness of an aluminium metal matrix composite, *Acta Metall. Mater.*, **38,** 489–96 (1990).
[6]S.P. Rawal, Metal-Matrix Composites for Space Applications (overview), *JOM,* **53,** 14-7 (2001).
[7]W S Lee and W.C. Sue, Dynamic impact and fracture behaviour of carbon fiber reinforced 7075 aluminum metal matrix composite, *J. Compos. Mater.*, **34,** 1821-41 (2000).

[8]D.R. Clarke, Interpenetrating phase composites, *J. Am. Ceram. Soc.*, **75,** 739–59 (1992).

[9]R. Soundararajan, G. Kuhn, R. Atisivan, S. Bose and A. Bandyopadhyay, Processing of Mullite–Aluminum Composites, *J. Am. Ceram. Soc.*, **84,** 509–13 (2001).

[10]J.W. Liu, Z.X. Zheng, J.M. Wang, Y.C. Wu, W.M. Tang and J. Lü, Pressureless infiltration of liquid aluminum alloy into SiC performs to form near-net-shape SiC/Al composites, *J. Alloy Compd.*, **465,** 239–43 (2008).

[11]H. Chang, R.L. Higginson and J.G.P. Binner, Microstructure and property characterization of 3-3 interpenetrating composites, *J. Mater. Sci.*, **45,** 662-8 (2010).

[12]J.G.P. Binner, H. Chang and R.L. Higginson, Processing of ceramic-metal interpenetrating composites, *J. Eur. Ceram. Soc.*, **29,** 837-842 (2009).

[13]A. Tasdemirci and I.W. Hall, The effects of plastic deformation on stress wave propagation in multi-layer materials, *Inter. J. Imp. Eng.*, **34,** 1797–813 (2007).

[14]D. Sherman and T. Ben-Shushan, Quasi-static impact damage in confined ceramic tiles, *Int. J. Imp. Eng.*, **21,** 245-265 (1998).

[15]H. Chang, J.G.P. Binner and R.L. Higginson, Ballistic evaluation of ceramic-faced, metal-ceramic interpenetrating composites, *Inter. J. Imp. Eng.*, submitted.

EFFECT OF AN INTERFACE ON DYNAMIC CRACK PROPAGATION

Hwun Park and Weinong Chen
School of Aeronautics and Astronautics, Purdue University
West Lafayette, Indiana, USA

ABSTRACT
    Dynamic crack propagation across an interface is investigated to understand the interaction between a propagating crack and an interface in glass under impact loading. Notched glass specimens having adhesive interfaces were impacted with plastic projectiles on the notches. Cracks developed from the notch tips and propagate into the interfaces perpendicularly. The patterns of crack propagations across the interfaces depended on the interface types. The crack stopped at the interface without adhesive. The crack passed across the interface with a very thin layer of adhesive. The crack branched into many cracks after it passed across the interface with an adhesive layer of finite thickness.

INTRODUCTION
    Glass has high potential for a transparent armor material because of its light weight, high hardness and transparency[1]. But glass has low tensile strength because of its surface micro-cracks, most of which are generated during manufacturing. Without such defects, glass shows great resistance to impacts. To effectively use glass in armor applications, the damage and failure behavior of glass under impact loading must be understood.
    The dynamic failure of brittle materials such as glass under a high speed impact is very complicated and successful models have not been developed. Glass has a unique dynamic failure phenomenon under high dynamic loading known as "failure wave". It is referred as crack networks propagating behind shock waves. Comminuted materials are created behind the failure wave, which affects the resistance to impacts. This phenomena has been explored in many studies but it is still not understood exactly[2,3]. The failure wave initiates at a glass surface and may stop at an interface of glass even though stress waves pass across the interface. After pausing at the interface, the failure wave may reinitiate from the interface[4]. It is very difficult to investigate the failure wave because it has too many cracks and associate variables. Studying single crack propagation first, and then extending to a general failure model is believed to be a viable way to investigate the phenomenon of failure wave.
    Layered structural design is widely used in armors. It has been known that the wave propagation is attenuated in the layered structures[5]. The different material properties of each layer cause stress waves to interact, result in stress attenuation. Moreover, the interface of each layer may arrest the failure propagation. It has been known that a crack may stop at the interface of two different media when it propagates in perpendicular to the interface. The static stress field of the crack tip ending in the interface was obtained with Airy function[6]. It was found that the rate of singularity and the principal stress direction depended on the properties of two media. Dally and Kobayashi studied crack arrests at the interfaces of duplex specimens with photoelastic methods[7]. In their experiments, a crack was initiated by static loading and stopped abruptly at the interface. The crack penetrated the interface when the load was high enough. The stress intensity factor decreased at the interface and increased sharply after the crack reinitiated at the second medium. Theocaris and Milios studied the crack propagation across the interface of bimaterial with the method of caustics[8]. A crack was initiated by a drop weight and penetrated the interface. The stress intensity factor versus time and crack extension showed different results from those of Dally and Kobayashi[7]. It increased before the interface, decreased abruptly at the interface and increased again. When the crack penetrated, it paused for tens of micro-seconds at the interface and reinitiated at the second medium. Moreover, the crack branched into multiple cracks in some cases. Xu and Rosakis investigated the crack propagation driven by low impact loading through interfaces of brittle polymer with photoelasticity[9]. The crack behavior

depends on the strength of adhesive. Cracks were arrested at the interface having low strength adhesive because two media were detached due to their weak bonding.

All those experiments were conducted under static or low impact loading conditions with brittle polymers such as Homalite. It is more difficult to investigate the dynamic fracture of glass because of its much faster crack speed and lower photoelastic constants. The crack propagation at an interface of glass driven by a dynamic load has not been investigated yet.

EXPERIMENT

This study aims at high impact loading with the impact velocity comparable to the crack velocity of glass, approximately 1500 m/s. To produce such impact velocities higher than 150 m/s, the projectile needs to be launched by a high speed gas gun. A 63.5 mm light gas was employed to produce consistent high velocity impact. The peak force and duration of pulse can be manipulated by adjusting the velocity and dimension of projectiles. Figure 1 shows the arrangement of the equipment. The gas gun discharges the projectile into the specimen. The couples of lasers and sensors detect the projectile. The velocity of projectile is obtained from the sensor signals. The high speed camera triggered by the laser and sensors begins to record the images of crack propagation at the instant when the projectile impacts the specimen. A bullet-proof polycarbonate shield and a mirror were employed to prevent the camera from any damage.

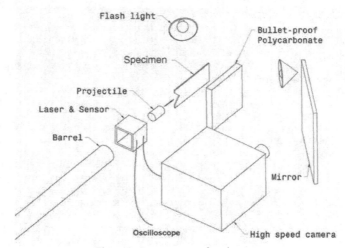

Figure 1. Arrangement of equipments

Figure 2 shows the dimensions of the projectile and specimen. The projectile strikes the notch and a crack initiates at the tip of the notch. The specimen is large enough to prevent a reflected tensile wave from the ends. Specimens were made with commercial soda-lime glass and Loctite E-30CL epoxy glass bond. The thickness of adhesive is controlled with shims and was measured with an optical magnifier. The projectile material requires low hardness and strength to prevent damage on the glass surface where the projectile initially contacts. Otherwise, the surface is subjected to developing fragmentations immediately when the projectile touches, because of the brittleness of glass. Smooth-On Featherlite™, casting polyurethane, was chosen for its very low hardness and light weight. High speed images and post-mortem observations verified that the cracks always initiate at the notch first instead of at the contact surfaces between the projectile and the specimen.

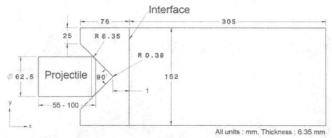

Figure 2. Dimensions of projectile and specimen

In regard to adhesive conditions, the thickness of adhesive is the only parameter varied in the experiments, because it controls wave interactions between two pieces of glass and the speed of crack propagation. The material properties of adhesive, the surface condition of adherends, and the adhesion strength are other important features that are not varied in the experiments reported here. The assumption was made that adhesive and adhesion are strong enough and the specimens do not have any cohesive and interfacial failure. The post-mortem inspection validated this assumption. All failures around the adhesive layer occurred on the glass but not the adhesive.

The high speed camera recorded the images of crack propagation and its interaction with the interface at a frame interval between 5 μs and 20 μs. Figure 3 shows the crack propagation on the specimen having two glass panels touching along the interface but without adhesive. Two plates were secured to touch each other without any bonding. The crack stopped at the interface. Similar phenomenon was observed in another study[9]. At that study, the crack arrested at the adhesive interface that was detached before the crack arrived. Besides the initial cracks, more cracks developed from the interface later, which were generated by the reflected tensile waves from the specimen edges.

Figure 3. Crack Propagation across an interface without adhesive (projectile: 217 g, 212 m/s)

Figure 4 shows the crack propagation across a very thin, less than 0.05 mm thick, adhesive layer. The crack passed across the interface without changing direction or branching. The propagation had a slight delay at the interface. A similar delay was also observed in another study[8].

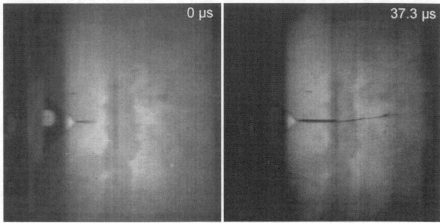

Figure 4. Crack Propagation across a thin interface (projectile: 162 g, 229 m/s)

Figure 5 shows the crack propagation across a 0.13-mm thick layer of adhesive. The crack branched into multiple cracks after it passed across the interface. The crack had a delay in crossing the interface, too. Considering that no branching happened in the interface with the very thin layer despite higher impact energy, this is an interesting phenomenon worth further exploration.

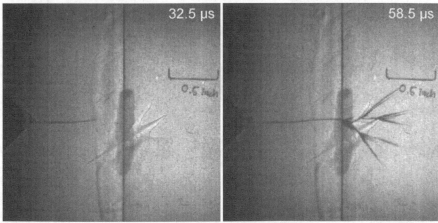

Figure 5. Crack propagation across a 0.13 mm-thick interface (projectile: 124 g, 192 m/s)

The specimen having a 1.3-mm thick layer of adhesive was impacted as shown in Figure 6. The crack had more branches and wider angles of branching than that happened in the specimen having the 0.13-mm thick layer as shown in Figure 5, in spite of similar impact energy. The thickness of the interface affected the pattern of the branching. The crack branched before reaching the interface in this experiment. The upper crack branch stopped at the interface while the lower branch penetrated the interface.

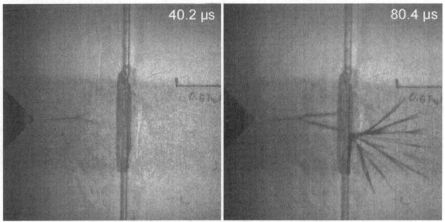

Figure 6. Crack propagation across a 1.3 mm-thick interface (projectile: 135 g, 200 m/s)

To obtain more quantitative information about the crack propagating in the glass plates, the method of caustics was employed to track the tip of cracks and to estimate stress intensity factors. The glass has very small photoelastic constants and the corresponding size of caustics is very small. K. Takahashi conducted optical test on tempered glass and measured the sizes of caustics[10]. They were only a few millimeters and it is hard to obtain the exact value of stress intensity factor of each crack with such small caustics. But it is possible to estimate the change of the stress intensity factor and compare the values of each branched cracks.

Figure 7 shows the configuration to observe the caustics of crack tips. The pulse laser and high speed camera are synchronized through the pulse generator to take instant pictures on the right time. Two mirrors were employed to put the laser and the camera away from the side of specimen, which prevent them from any damage.

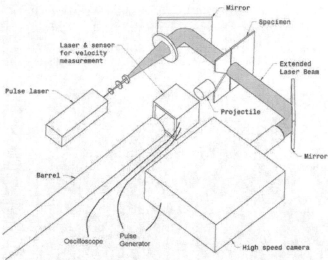

Figure 7. Equipments configuration for optical investigation

The diameter of the pulse laser beam is only 0.6 mm which was extended to cover the region where the crack propagates through. Figure 8 shows the dimensions and arrangement of lenses to extend the beam diameter by 100 times.

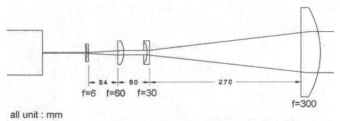

Figure 8. Dimension of 100 times laser beam extender

Figure 9 shows the image of caustics. The vertical black line is the interface having a 0.5-mm thick adhesive layer. The horizontal black line is a shadow that is not covered by laser pulse. The size of caustics is approximately 1 mm. As shown in the figure, the initial crack was branched before it reached the interface, which is similar to that of Figure 6. Crack 1 has the higher stress intensity factor than crack 2 initially. But when they reach at the interface, crack 2 turns out to have a higher stress intensity factor; and it penetrates the interface and branches in to multiple cracks. This phenomenon indicates that the main crack driving force may switch from one crack to another when the cracks approach an interface delaying their crack propagation. It explains why one crack penetrates the interface while other crack stops at the interface even though they are branched from a single crack and connected each other as shown in Figure 6.

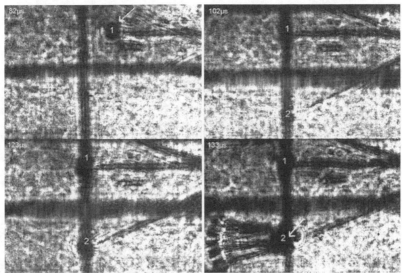

Figure 9. Caustic of crack propagation across a 0.5 mm-thick interface (projectile: 84 g, 262 m/s)

The fracture surfaces generated by crack propagation were inspected visually. Figure 10 shows the specimen having the very thin layer of adhesive which was retrieved from debris. The fracture surface in Figure 10 B shows the transition of the roughness of surface generated by crack branching. The roughness in the glass specimen in the front of and at the back of the interface may reveal information on the fracture energy. The fracture surface is mirror-like smooth and roughness does not change when the crack passes across the interface. This is not surprising because the interface has only a very thin layer of adhesive and the crack has barely interfered with it.

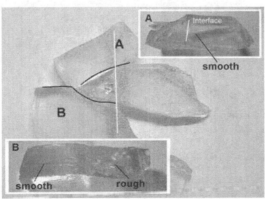

Figure 10. Fracture surface roughness at thin interface

DISCUSSIONS

A slight time delay was observed when a crack passed across the layer of adhesive. The delay time depends on the thickness of adhesive layer that has a much lower crack velocity than that in glass. While crossing the very thin interface, it took remarkable time before the crack branched. It is known that a crack does not initiate immediately even after stress waves reach and needs some time to gain enough energy to develop itself[11]. It is considered that the slow crack speed in epoxy adhesive and the time to accumulate the energy to initiate cracks in the second medium caused the delay.

The crack branched into multiple cracks if the interface had a layer of finite thick adhesive. As the thickness increased, the number of cracks was observed to increase. The mechanism of crack branching have not been explained clearly. The review by Ramulu and Kobayashi described that there had been inconsistency in the observations of crack branching[12]. The velocity may be constant or decrease after the bifurcation. The definition of the critical conditions for crack bifurcation have been tried in various ways such as a critical velocity, a stress intensity factor, a strain release energy, a surface roughness, and so on. None of them show a successful correlation with experimental results. Ravi-Chandar reported that the stress intensity factor of each crack decreased after bifurcation and the branch angles depended on the stress wave interaction at the crack front[13].

There are two possibilities for crack branching at the interfaces having finite thick layers. First, the delayed cracks accumulate more fracture energy and this excessive energy causes immediate crack branching at the second medium. Second, the crack has a larger plastic zone at the interface with a ductile adhesive layer and has larger stress distribution at the crack tip sufficient to cause crack branching. Further experiments are needed to find the exact reasons.

The roughness of fracture surface is an indication of excessive energy because the energy dissipates through forming new surfaces[12,13]. It is known that the roughness of the fracture surface changed into mirror, mist and hackle as the crack propagates[12]. Figure 10 B shows that the roughness increased with crack branching as reported by other studies. However, the roughness did not change as the crack pass across the interface with a very thin layer. This means that the fracture energy did not change in the crossing.

CONCLUSIONS

Dynamic crack propagation across an interface in a glass plate driven by high dynamic loading was investigated. The thickness of adhesive layer affects the patterns of the crack propagation across the interface. The crack penetrates the interface with a very thin layer of adhesive and the fracture energy does not change while crossing. The crack branches into multiple cracks if the layer has a finite thickness. The crack stops at the interface without any adhesion. The stress intensity factor of each branched crack may increase or decrease when the cracks approach the interface.

REFERENCES

[1] X. Nie, W. Chen, X. Sun, D. W. Templeton, Dynamic Failure of Borosilicate Glass under Compression/Shear Loading Experiments, *J. Am. Ceram. Soc.*, 2007, **90**, 2256-62 (2007).
[2] G. I. Kanel, S. V. Razorenov, V. E. Fortov, Shock-wave Phenomena and the Properties of Condensed Matter, Springer (2004).
[3] R. Feng, Formation and Propagation of Failure in Shocked Glass, *J. App. Phy.*, **87**, 1693-1700 (2000).
[4] G. I. Kanel, A. A. Bogatch, S. V. Razorenov, Z. Chen, Transformation of Shock Compression Pulses in Glass due to the Failure Wave Phenomena, *J. App. Phy.*, **92**, 5045-52 (2002).
[5] Y. Oved, G. E. Luttwak, Z. Rosenberg, Shock Wave Propagation in Layered Composites, *J. Comp. Mat.*, **12**, 84-96 (1978).

[6.] A. R. Zak, M. L. Williams, Crack Point Stress Singularities at a Bi-Material Interface, *Tran. ASME*, **3**, 142-143 (1963).

[7.] J. W. Dally, T. Kobayashi, Crack Arrest in Duplex Specimens, *Int. J. Sol. Struc.*, **14**, 121-129 (1977).

[8.] P. S. Theocaris, J. Milios, Crack-arrest at a Bimaterial Interface, *Int. J. Sol. Struc.*, **17**, 217-230 (1981).

[9.] L. R. Xu, A. J. Rosakis, An Experimental Study of Impact-involved Failure Events in Homogeneous Layered Materials Using Dynamic Photoelasticity and High-speed Photography, *Opt. Laser. Eng.*, **40**, 263-288 (2003).

[10.] K. Takahashi, Fast Fracture in Tempered Glass, *Key Eng. Mat.*, **166**, 9-16 (1999).

[11.] H. Schardin, Velocity Effects in Fracture, *Fracture : Proc. Int. Conf. Atomic. Mech. Frac. Help in Swampscott*, Massachusetts, 297-330 (1959).

[12.] M. Ramulu, A. S. Kobayashi, Mechanics of Crack Curving and Branching – a Dynamic Fracture Analysis, *Int. J. Frac.*, **27**,187-201 (1985).

[13.] K. Ravi-Chandar, W. G. Knauss, An Experimental Investigation into Dynamic Fracture: III. On Steady-state Crack Propagation and Crack Branching, *Int. J. Frac.*, **26**, 141-154 (1984).

DYNAMIC EQUATION OF STATE AND STRENGTH OF BORON CARBIDE

Dennis E. Grady
Applied Research Associates, Southwest Division, 4300 San Mateo Blvd NE
Albuquerque, New Mexico 87110, USA

ABSTRACT

Boron carbide ceramics have been particularly problematic in attempts to develop adequate constitutive model descriptions for purposes of computational simulation of dynamic response in the ballistic environment. Dynamic strength properties of boron carbide ceramic differ uniquely from comparable ceramics, and are difficult to characterize. Further, boron carbide is suspected of undergoing polymorphic phase transformation within the shock-wave compression process. These phase transformation features have been particularly elusive under experimental investigation, and consequently are also difficult to capture within an appropriate constitutive model. In the present paper shock-wave compression measurements conducted over the past forty years are reassessed for the purposes of achieving improved understanding of the dynamic response of boron carbide. In particular, attention is focused on the often ignored Los Alamos National Laboratory (LANL) Hugoniot measurements performed on porous sintered boron carbide ceramic. The LANL data exhibit two compression anomalies on the shock Hugoniot within the range of 0-60 GPa that may relate to crystallographic structure transitions. Recent molecular dynamics simulations on the compressibility of the boron carbide crystal lattice reveal compression transitions that bear intriguing similarities to the LANL Hugoniot results. Contemporary and later Hugoniot measurements, however, performed on near full density boron carbide ceramic differ starkly from the LANL Hugoniot data. Comparisons in a shock velocity versus particle velocity representation of the data that normalize porous compaction and precursor strength effects are particularly revealing. These other data exhibit markedly less compressibility and tend not to show comparable anomalies in compressibility. Experimental uncertainty, Hugoniot strength, and phase transformation physics are explored as explanations for the marked discrepancy. It is argued that experimental uncertainty and Hugoniot strength are not likely explanations for the observed differences. The notable mechanistic difference in the processes of shock compression between the LANL data and that of the other studies is the markedly larger inelastic deformation and dissipation experienced in the shock event due to compaction of the substantially larger porosity in the LANL test ceramics. High-pressure diamond anvil cell experiments reveal extensive amorphization, presumed to be a reversion product of a higher-pressure crystallographic phase, that is a consequence of application of both high pressure and shear deformation to the boron carbide crystal structure. The dependence of shock-induced high-pressure phase transformation in boron carbide on the extent of shear deformation experienced in the shock process, offers a plausible explanation for the differences observed in the LANL Hugoniot data on porous ceramic and that of other shock data on near-full-density boron carbide.

INTRODUCTION
Properties of boron carbide ceramic relating to dynamic equation of state and strength, despite extensive studies of the material directed towards these properties, have been particularly elusive. Consequently, physical understanding of, and computational models that describe, the response of boron carbide in the impact and ballistic environment are not yet adequate.

Boron carbide ceramic is, nevertheless, an attractive candidate for ballistic applications. It is the lowest density and the highest shock strength ceramic among a suite of ceramics under consideration for such applications. As noted, many of the necessary physical, mechanical and crystal properties of boron carbide have been widely studied and access into much of the literature is available in the reports of Emin[1], Aselage et al.[2] and Dandekar[3]. In particular, the Hugoniot and ultrasonic data integral to development of shock-wave and high-strain-rate constitutive response models are reviewed in the latter report. More recent papers, also detailing much of the earlier Hugoniot work along with additional newer, high-resolution shock wave data and analysis, are that of Vogler et al.[4], Zhang et al.[5], Holmquist and Johnson[6], and Ciezak and Dandekar[7].

The primary focus of the present paper is on the dynamic equation-of-state (EOS) and shock strength properties of boron carbide ceramic. Namely, the dynamic compressibility and strength of the material over the shock pressure range of concern to ballistic applications (approximately 0-60 GPa). Previous shock EOS studies[8] suggest dynamic compressibility is complicated by polymorphic phase transformation and subsequent research have supported, at least in part, such conjecture. Despite growing evidence, the existence and nature of phase transformation in boron carbide within the shock-wave environment remains obscure.

In this paper EOS compressibility, high-pressure phase transformation and dynamic strength of boron carbide ceramic in the shock environment are examined through detailed analysis of shock Hugoniot data. Although other methods are being explored to characterize compressibility and high pressure phase transformation in ceramic materials – diamond anvil cell technology for one – shock compression techniques remain the standard when issues of phase transition pressures and compressibility are of paramount concern. Extraction of the required phase transformation properties from shock wave data remains elusive, however, and usually requires concerted effort by workers experienced in the field.

The extensive shock Hugoniot data for boron carbide of McQueen et al.[9], provided in the LANL shock compendium[10] is the principal source of data used in this study. (This landmark paper was jointly co-authored by Robert McQueen, Stanley Marsh, John Taylor, Joe Fritz and William Carter, all from what is now Los Alamos National Laboratory. Since this seminal paper, and the boron carbide Hugoniot data, will be referred to many times throughout this report, the reference will just be McQueen with no intention of minimizing the substantial contributions of the co-authors.) These data, joined with the necessary analysis, provide strong evidence for phase transformation in boron carbide ceramic in addition to the compressibility properties accompanying the phase transformation. Some of this analysis effort has been documented in earlier project reports[11,12].

Recently, observations of amorphous regions in boron carbide subjected to intense shock through ballistic impact have been reported by Chen et al.[13]. Amorphization from shock compression can be a signature of unique high-pressure phases. This issue has been explored through diamond anvil cell (DAC) studies of pressure-induced amorphization of boron carbide and through supporting molecular dynamics (MD) simulations by Yan et al.[14]. Intriguing consistencies among the shock equation-of-state features, DAC observations, and MD simulations offer hope that improved understanding of boron carbide physics under intense dynamic loading is emerging.

A significant setback occurs, however, when the broader base of shock Hugoniot data is examined. These include the studies of Wilkins[15], Gust and Royce[16], Pavlovski[17], Vogler et al.[4] and Zhang et al.[5]; all on near-full-density boron carbide ceramics. (sintered boron carbide ceramic samples used in the Hugoniot tests of McQueen all had substantial porosity.) These other sets of Hugoniot data are each independently consistent and exhibit results that differ significantly from that of McQueen and the several Hugoniot points of Grady[8]. Possible causes for the differences due to dynamic strength, phase transformation, material variations and experimental error are examined in the present study. From this effort a level of understanding is achieved, although some of the discrepancies in the shock-wave data for boron carbide ceramic continue unresolved.

The dynamic or shock strength of ceramic in the ballistic environment is a key aspect of the constitutive model necessary for computational analysis. Shock data are commonly used to achieve at least a first-order estimate of strength. An assessment of the initial dynamic strength at failure of the material is provided by the Hugoniot elastic limit (HEL), which is determined in turn from the amplitude of an elastic precursor wave from a structured shock wave measurement. Post HEL dynamic strength can be independently assessed with shock-wave data alone through several methods[3,4,18]. Alternatively, comparative use of static high-pressure hydrodynamic compression data and shock Hugoniot measurements can be used to determine dynamic strength at pressure[19]. Elastic precursor wave measurements provide quite high initial strength (15-18 GPa shock pressure) for boron carbide ceramic; although, unique stress relaxing and chaotic precursor wave structure are suggestive of a time-dependent and heterogeneous character to dynamic shear failure. The transient loss and subsequent recovery of strength in the shock compression event are indicated in more recent studies[4,18]. The interplay of possible phase transformation and dynamic strength remains poorly understood.

HUGONIOT EQUATION OF STATE OF MCQUEEN

Most analysts that have attempted to construct a constitutive model for the shock properties of boron carbide ceramic have by-passed the shock Hugoniot data of McQueen et al.[9] (see Marsh[10]) in favor of concurrent and later shock data. This is because the McQueen study was performed on ceramics with significant levels of porosity, whereas most other studies were performed on near-full-density material. The investigation of Dandekar[3] on the equation of state (EOS) and strength of boron carbide makes use of Hugoniot data of McQueen from some of the lower porosity material but not that of the higher porosity material. Actually, when assessing the high-pressure shock equation of state of ceramic materials, a degree of porosity in the material has some attractive features. The high dynamic strength and modest Poisson's ratio of boron carbide leads to a shock transition to the Hugoniot state through a double shock wave. The first precursor wave carries the ceramic to the Hugoniot elastic limit (HEL). The second, to the final Hugoniot state. The structured VISAR profile in Fig. 1 is representative[8], exhibiting an HEL of approximately 18 GPa and a final state Hugoniot pressure in the neighborhood of 40 GPa. This two-step transition to the final Hugoniot state occurs in fully-dense boron carbide until shock pressures well in excess of 100 GPa are achieved. Although not a significant complication, accounting for the elastic precursor and HEL state adds one further step, and added uncertainty, in the analysis providing the final shock Hugoniot states.

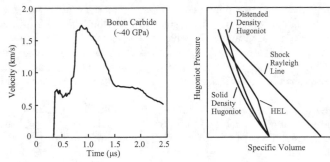

Figure 1. Structured shock wave profile from VISAR measurements of Grady[8]. The corresponding Shock Hugoniot and two-step shock load path are shown on the right. Single shock compression of porous solid to a distended density Hugoniot state is shown for comparison. (The lower pressure crush Hugoniot is not illustrated.)

In contrast, with distended (porous) ceramics with sufficient levels of porosity, transition to the Hugoniot state can be achieved through a single shock. Actually, a sintered porous ceramic will exhibit a modest HEL, however it is overdriven at much lower pressures and is a comparably smaller contribution to the Hugoniot analysis and uncertainties over much of the pressure range. McQueen tested boron carbide with two porosities: one closer to full density (3-7% porosity) and one quite porous (22-25% porosity). The amplitude dependence of the HEL on porosity for sintered boron carbide ceramic has been studied by Brar et al.[20]. Representative Hugoniot states and the Rayleigh line path for the initially distended (or porous) material are illustrated on the right in Fig. 1.

A further feature of the shock compression of a porous ceramic is the extensive deformation and dissipation associated with the crush and compaction of the void space. Two issues are of importance: First, the compaction process entails extensive non-hydrostatic stress and inelastic shear on the microscale well above that incurred in shock compression to the same pressure on a fully dense ceramic. Second, due to the excessive deformation, dissipation temperatures, both heterogeneous and on average, are achieved that are substantially higher than comparable shock compression of the fully dense ceramic. Both issues are addressed further. At this point it is worth noting that either incomplete pore compaction, or excessive temperature rise, would lead to shock states that are at lower density than would be the case if either effect did not occur. The higher temperature or incomplete compaction Hugoniot is suggested in the plot in Fig. 1.

Again, the principal shock-wave data for boron carbide ceramic analyzed in this section are that of McQueen provided in the Los Alamos Compendium[10]. As noted, other shock data on boron carbide are available[4,5,8,15,16,17], and will be considered shortly, but the former data have unique features significant to the present objectives. Briefly, analysis of the shock-wave data of McQueen provides a unique description of the shock equation of state of boron carbide. These shock data alone strongly suggest the occurrence of one phase transformation in the shock compression process. Shock data combined with extrapolation of modest-pressure ultrasonic data suggest the occurrence of a second transition under shock compression. These efforts have been previously detailed in project reports[11,12].

Before proceeding on this facet of the study, however, some qualifying remarks concerning the broader available Hugoniot data for boron carbide are necessary. As pointed out in the introduction, there is a marked disparity between the shock Hugoniot compression data of Wilkins[15], Gust and Royce[16], Pavlovsky[17], Vogler et al.[4] and Zhang et al.[5] on one hand, and the data of McQueen et al.[9] and Grady[8] on the other. Whether these differences are due to manifestations of material strength, effects of phase transformation, uncertainties in initial material composition and chemistry, or difficulties with analysis and interpretation of the measured shock data is not yet fully understood. Some of the issues have been discussed by Dandekar[3]. These issues are central to the present study and will be addressed further in later sections of the present paper.

The McQueen data provide shock compression states for boron carbide over the pressure range of about 15–115 GPa. These data encompass two starting densities for boron carbide of nominally 2400 kg/m^3 and 1900 kg/m^3. Assuming reasonably pure material, the sample-to-sample reported densities imply initial porosities of approximately 3-7% and 22-25%, respectively, based on a theoretical density of 2520 kg/m^3. (As ceramic composition is not reported, any sintering additives would make the calculated porosity of boron carbide approximate. Appreciable higher density additives, and correspondingly higher calculated porosity, could account in part for discrepancies with later Hugoniot data. Excess carbon would have the converse effect.) Shock velocity versus particle velocity ($U-u$) and pressure versus specific volume ($p-v$) Hugoniot data for the boron carbide are shown in Fig. 2. In the test method of McQueen shock velocity and particle velocity are the measured quantities and are identified as the original data in the right hand plot. There are two regimes associated with the two initial porosity levels. The scatter for the most part is due to sample-to-sample variations in density (and porosity).

Pressure and specific volume are calculated from the measured shock and particle velocity data through the Hugoniot relations,

$$p = \rho_{oo} U u, \qquad v = (1 - u/U)/\rho_{oo}. \tag{1}$$

with ρ_{oo} being the initial distended (or bulk) sample density. The $p-v$ Hugoniot for all of the McQueen data is shown on the left in Fig. 2 and exhibits some interesting behavior. As alluded to previously, shock compaction of the porous ceramic will lead to excessive dissipative heating – substantially more so for the higher 22-25% porous material than for the lower 3-7% porous material. The extent of heating, however, is not sufficient to separate the two sets of data in the pressure-volume plot through thermal pressure (Gruneisen) properties of the material. Where Hugoniot states for the two porosities overlap, pressure-volume values do not statistically differ. On the other hand, heating due to pore compaction is expected to markedly reduce residual strength at the shock state leaving material closer to hydrodynamic conditions (unable to support significant deviator stress). Strength states on the Hugoniot for near-full-density ceramic is an issue of interest and has been pursued for boron carbide in the work of both Dandekar[3] and Vogler et al.[4]. The lowest three Hugoniot points for the lower porosity ceramic of McQueen appear to exhibit some strength features on the Hugoniot, although it is known that the test method of McQueen is less reliable at the lower shock amplitudes where structured shock waves are probably present.

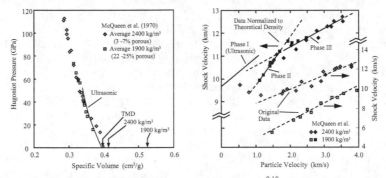

Figure 2. Shock Hugoniot data for boron carbide ceramic[9,10]. On the right are shown both the experimental and normalized shock velocity versus particle velocity data. Extrapolated compressibility from the ultrasonic study of Manghnani et al.[21] is also shown with the Hugoniot data. Three phases of boron carbide inferred from the shock data are identified (see text).

The last observation emphasizes a point elaborated on previously. Shock Hugoniot states achieved on near-full-density ceramic, which includes the work of Gust and Royce[16], Grady[8] and Vogler et al.[4], do so through a structured shock wave. Hugoniot states are referenced from the Hugoniot elastic limit (HEL) state that is itself determined from the magnitude of a rather unusual, and probably unsteady, elastic precursor wave[4,8]. Although not eliminated, any effect of the precursor wave transition to the final Hugoniot state is mitigated by the porosity in the McQueen materials. This important difference may relate to some of the disparity in the high-pressure Hugoniot data for boron carbide ceramic.

Synthesis of the two sets of the McQueen Hugoniot data is capitalized on to assess high pressure equation of state features of the boron carbide. The pressure versus specific volume Hugoniot data on the left in Fig. 2 are mapped back into the more revealing shock velocity versus particle velocity plane through the Hugoniot relations,

$$U' = \sqrt{\frac{p}{\rho_o(1-\rho_o\upsilon)}}, \qquad u' = \sqrt{\frac{p(1-\rho_o\upsilon)}{\rho_o}}. \tag{2}$$

where $\rho_o = 2520$ kg/m^3 is the theoretical density of boron carbide. (Note the use of bulk density in Eqns. (1) and solid density in Eqns. (2).) These shock velocity versus particle velocity states, referenced from the initial theoretical density of boron carbide, provide a sensitive display of compressibility features on the shock Hugoniot. This presentation of the Hugoniot data is identified by the solid symbols in Fig. 2. This particular presentation of the Hugoniot data is referred to as the normalized shock velocity versus particle velocity representation and is used later in the paper with Hugoniot data of other workers. When used with caution, this display of the data can reveal features of the Hugoniot compression not readily observed in other plots.

With this normalization of the shock data for the widely different initial porosity test samples, the graph on the right in Fig. 2 displays clear evidence for bi-linear behavior in the shock velocity versus particle velocity plot, suggestive of a shock-induced phase change on the principal Hugoniot. The bi-linear curve shown in Fig. 2 is the result of a statistical analysis of the data to determine linear shock versus particle velocity constants for boron carbide in the two regions identified as shock phase II and shock phase III in Table 1 and in the figure[11]. All fits to the data provide a pressure intercept between 45 and 55 GPa (the shock transition pressure). The optimum values of C_o (the intercept) and S (the slope) for the linear fits to the regions identified as phase II and phase III in Fig. 2 intercept closer to 45 GPa and determine the 45-50 GPa transition pressure identified in Table 1.

Table 1. Shock phase transformation properties of boron carbide inferred from the shock Hugoniot data of McQueen et al.[9,10] and ultrasonic measurements of Manghnani et al.[21].

Phase	C_o (km/s)	S	p_t (GPa)	$\Delta\upsilon_t$ (cm^3/g)
B$_4$C-I	9660	1.32	Ambient Phase	Ambient Phase
B$_4$C-II	7700	2.15	< 27	≈ 0.006
B$_4$C-III	9800	0.80	45–50	< 0.002

The shock data is unable to distinguish between a second order transition (no transition volume change) and a small first order transition. A recent report of a shock transition in B$_4$C at 38 GPa by Zhang et al.[5] with second order transition characteristics is possibly the same inflection in the McQueen data identified here. Vogler et al.[4] report a similar transition indication from shock data near 40 GPa. The high pressure slope in phase III of $S = 0.8$ cannot, of course, persist as the asymptotic limit implied by the linear U vs. u Hugoniot is unphysical.

A plot of the McQueen Hugoniot data below 60 GPa is shown in Fig. 3. The overlying curve is the linear $U - u$ Hugoniot provided by C_o and S for phase II in Table 1. As argued previously, the curve should provide a sensible estimate of the hydrodynamic compressibility of boron carbide within the stated range. There are some additional arguments for this conclusion. First, if there were a significant strength offset to the Hugoniot stress it would be highly fortuitous for the stress offset for the markedly different starting density materials to be the same. The argument would also support full compaction of the pore volume by the initial shock wave, and further suggests that local shear stresses are small compared to the compressive shock pressure. Second, and more convincing, elevated temperatures at the Hugoniot states for the more porous materials are inconsistent with significant material strength. In the range of 40 GPa shock pressure, containing a nice cluster of both 1900 kg/m^3 and 2400 kg/m^3 material Hugoniot points, temperatures of the lower initial density material subjected to Hugoniot shock compression are readily estimated. Based on a constant specific heat of 960 J/kg K (Dandekar[3]), an ambient (post-shock) average temperature rise due to shock heating is calculated to be about 2700 K. This level of temperature rise would approach the ambient melt temperature of boron carbide which is also about 2700 K (Dandekar[3]). The specific heat can be expected to increase with temperature and therefore the calculated temperature rise is an overestimate. Nonetheless, it is unlikely that such temperature states on the Hugoniot would support significant material strength. Further, due to low thermal conductivity, it is also unlikely that thermal equilibrium on the microscale is achieved, leading to locally hot deformation regions that would additionally reduce material strength on the Hugoniot.

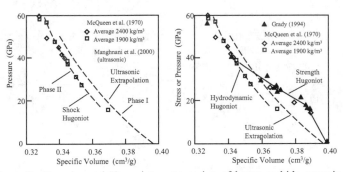

Figure 3. Experimental Hugoniot compression of boron carbide ceramic and extrapolation of the ultrasonic compressibility of boron carbide. Hugoniot states exhibiting clear strength effects are included on the right.

In the next section the application of low-pressure ultrasonic studies to assess the higher-pressure compressibility properties of boron carbide is described (Manghnani et al.[21]). The compressibility of boron carbide resulting from this experimental investigation is compared with the shock Hugoniot data in Figs. 2 and 3. The emphasis here is the comparison on the left in Fig. 3 with the lower-pressure Hugoniot states of McQueen for the higher porosity material. At a Hugoniot pressure of 16 GPa, and again at 27 and 31 GPa, Hugoniot compressions are markedly larger than the ultrasonic compression curve. The single Hugoniot point at 16 GPa can certainly be questioned. It is, however, difficult to assign uncertainties either to the shock Hugoniot measurements, or to the ultrasonic extrapolation, sufficient to account for the differences at higher pressures. This comparison anomaly is here identified as a Hugoniot phase transition occurring at a pressure below 27 GPa and with a volume strain, determined from the disparity between the Hugoniot and ultrasonic compression curves, of approximately 0.006 cm^3/g. Speculation as to its nature will wait until later in the paper. The ambient phase is identified in Table 1 with C_o and S provided by the ultrasonic measurements.

Comparisons with the ultrasonic extrapolation are also shown in Fig. 2 where Hugoniot phases I, II and III are identified in the shock velocity versus particle velocity plot.

ULTRASONIC EQUATION OF STATE OF BORON CARBIDE

The extrapolation of lower-pressure ultrasonic elastic properties of solids to high pressures can provide a useful reference for the evaluation of shock Hugoniot data. Both dynamic strength and shock-induced change of state are frequently revealed through such comparisons.

Ultrasonic properties of boron carbide ceramic as a function of hydrostatic pressure to 2.1 GPa have been measured by Manghnani et al.[21]. The material tested was a high-quality, isotropic, full-density polycrystalline boron carbide provided by Dow Chemical Co. Reported initial density was 2527 kg/m^3. The theoretical density of boron carbide is 2520 kg/m^3.

Ambient ultrasonic longitudinal and shear wave speeds measured for the polycrystalline ceramic are 14.03 km/s and 8.84 km/s, respectively. Calculated initial elastic properties are 234.9 GPa, 197.3 GPa and 0.17 for the Bulk modulus K, the shear modulus G and the Poisson's ratio ν, respectively. Linear increase in the longitudinal and shear frequency ratios to 2.1 GPa provide isentropic pressure derivatives for the bulk modulus of $K' = 4.26$ and shear modulus of $G' = 1.10$. The results of Manghnani et al.[21] can be compared with values of $K_o = 240$ GPa and $K'_o = 4.67$ for the bulk modulus and first pressure derivative determined by Dodd et al.[22] on boron carbide ceramic from Cercom Co. with a density of 2514 kg/m^3 determined from ultrasonic velocities to 0.2 GPa.

A sensible extrapolation of the compressibility to higher pressures is provide by the Birch finite-strain relation[23] which has the functional form,

$$p(\upsilon) = \frac{3}{2} K_o \left(\frac{\upsilon}{\upsilon_o}\right)^{-\frac{7}{3}} \left[1 - \left(\frac{\upsilon}{\upsilon_o}\right)^{\frac{2}{3}}\right] \left[1 + \frac{3}{4}(K'_o - 4)\left(\left(\frac{\upsilon}{\upsilon_o}\right)^{-\frac{2}{3}} - 1\right)\right]. \tag{3}$$

The Birch equation and the compressibility relation resulting from the linear relation between shock velocity and particle velocity $U = C_o + Su$,

$$p(\upsilon) = \frac{\rho_o C_o^2 (1 - \upsilon/\upsilon_o)}{\left[1 - S(1 - \upsilon/\upsilon_o)\right]^2}. \tag{4}$$

are second-order coincident if $K_o = \rho_o C_o^2$ and $K' = 4S - 1$ are satisfied[24].

The ultrasonic equation of state for boron carbide provided by Manghnani et al.[21] and extrapolated by the Birch finite strain equation is compared with the Hugoniot data of McQueen in Figs. 2 and 3. The data of McQueen from the lowest point at 16 GPa and higher reside on a markedly higher density Hugoniot path than is suggested for the compressibility of boron carbide from the ultrasonic extrapolation. The second order compressibility term K'' is not known; however the measurements of Manghnani et al. to 2.1 GPa show $K'' \cong 0$ within measurement uncertainty. It is unlikely that a finite K'' could account for the disparity. A reasonable explanation is shock-induced phase transformation in the lower range of the shock Hugoniot data.

The extrapolated compressibility of boron carbide based on the ultrasonic measurements for full density polycrystalline boron carbide[21] is somewhat stiffer than the calculated molecular dynamics simulations[14] of the hydrostatic compressibility of a unit cell. The trend of this difference is consistent with the constraints imposed by polycrystallinity where the crystal grains have anisotropic elastic properties.

STRUCTURED SHOCK-WAVE MEASUREMENTS ON BORON CARBIDE

In the mid 1990's shock-wave measurements were performed on a high-quality full-density ceramic using time-resolved velocity interferometer (VISAR) instrumentation[8]. Comparable structured wave measurements using VISAR were performed in a later study of boron carbide shock-wave properties[4]. This later work will be described in a following section. A representative structured wave profile from the earlier study is shown in Fig. 1.

The boron carbide ceramic studied in the earlier investigation was prepared and provided by Dow Chemical company, and is nominally the same ceramic tested in the ultrasonic study of Manghnani et al.[21]. Density of the ceramic was independently measured to be 2506 kg/m³, which again can be compared with the theoretical density of 2520 kg/m³ for boron carbide. The density is slightly lower than that reported by Manghnani et al., but within 1%. Nominal grain size was 3 μm. Uniform and consistent ultrasonic velocities of 14.07 km/s and 8.87 km/s were also independently measured for the longitudinal and shear velocities, respectively; again, in good agreement with Manghnani et al. These velocities provide elastic properties of 233.4 GPa, 197.2 GPa and 0.17 for the bulk modulus, shear modulus and Poisson's ratio, respectively.

Hugoniot properties determined from the VISAR structured wave measurements were initially reported by Grady[8]. The same data were later reanalyzed using a different method referred to as the pulse-echo technique[12]. The latter technique is judged to be the more accurate when differences are observed in the two analysis methods. Such differences occur when the second deformation shock wave achieves states that are only modestly higher than the Hugoniot elastic limit (HEL). In the shock wave experiment the time history of the structured wave profile is measured at the interface between the boron carbide sample and a transparent lithium fluoride crystal. The latter is a laser window material that has been calibrated for application in the VISAR diagnostic method of Barker and Hollenbach[25].

The conditions of the experiment are illustrated in Fig. 4. A projectile carrying a disc of boron carbide undergoes planar impact onto a stationary thicker disc of boron carbide. The impact creates a one-dimensional compressive shock wave that propagates through the stationary ceramic sample. The wave separates into a faster elastic precursor wave and a following deformation shock wave compressing the material to the final Hugoniot stress and strain state. The material is subsequently decompressed through a following release wave originating at the back free surface of the projectile ceramic disc.

The transmitted shock and release waves emerge at the ceramic surface opposite impact and accelerates this surface. Velocity interferometry (VISAR) is used to measure the velocity history of this surface imparted by the transmitted shock pulse. In the present test the transparent lithium fluoride (LiF) window is mounted on the sample as shown. Velocity measurement is made at the interface between the sample and LiF window. Issues of concern here are equally present if velocity measurement is made at a free surface.

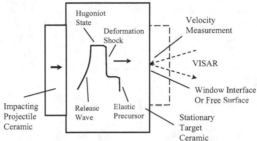

Figure 4. Illustrates features of the structured shock wave experiment with VISAR time-resolved velocity diagnostics.

The measured velocity history, from which Hugoniot properties of the material are inferred, is complicated by the two-wave structure of the compressive shock wave. The precursor wave arrives at the measurement interface first, reflects, and subsequently reverberates between this interface and the following deformation shock wave as illustrated on the right in Fig. 5. This complication seriously distorts the recorded velocity profile from the unblemished structured shock wave before interference with reflections from the interface. This distortion confuses the arrival of the shock wave necessary to the Hugoniot state calculation.

The plot on the left in Fig. 5 shows one experimental profile in boron carbide illustrating arrival of the elastic precursor (point 1), the first reverberation of the precursor (point 2), and the subsequent arrival of the deformation shock wave (point 3). Commonly the latter point is used as an estimate of the deformation shock arrival and hence the shock velocity and this was done in the earlier analysis of the present boron carbide data[8] . These previous Hugoniot states for boron carbide interpreted from the velocity measurement (transmitted shock method) are plotted in Fig. 6. As shown in the distance-time plot in Fig. 5, this method errors by not accounting for the reduction in shock speed caused by subsequent reflections of the precursor wave.

An alternative method for determining the deformation shock velocity from the interface velocity profile can be determined from the first arrival (point 1) and first reflection (point 2) of the elastic precursor wave. With reasonable estimates of the precursor wave velocity the position and time at which the following shock wave is first perturbed by the reflected precursor wave can be determined and, consequently, also the shock velocity. Hugoniot states determined by this precursor echo method are also shown in Fig. 6.

This latter method (pulse echo) is believed to be the more accurate, particularly at the lower amplitudes at which the precursor is a significant fraction of the total shock amplitude. As observed in the comparison, discrepancies are observed in the Hugoniot points up to approximately 30 GPa. Differences are not discernable in the higher 40 GPa and 56 GPa points. A discontinuity is observed in the shock versus particle velocity data between approximately 30 and 40 GPa shock pressure. The McQueen Hugoniot data suggest a possible transition from a tentative B4C-I to the B4C-II phase somewhere below 27 GPa shock pressure. The present Hugoniot data suggest a compression anomaly in the pressure range of 30–35 GPa. Any significance in this discontinuity is mitigated by the sparse data and difficulties of data analysis in this region. Hugoniot elastic limit and Hugoniot states for the present data are compared with that of McQueen on the right in Fig. 3.

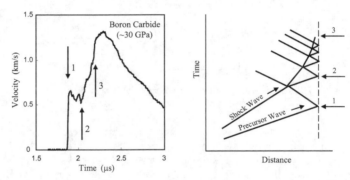

Figure 5. A VISAR velocity history measurement at the interface between boron carbide test sample and lithium fluoride laser window[8]. Comparable wave features are identified on the right and left figures due to wave reflection at the interface from wave impedance differences.

HIGH-PRESSURE AMORPHIZATION OF B$_4$C

The shock Hugoniot data of McQueen supports the possibility of high-pressure phase transformation of boron carbide in the shock environment. Regions of amorphous B$_4$C have been observed in boron carbide ceramic subjected to shock through ballistic impact by Chen *et al.*[13]. Amorphization of boron carbide subjected to extreme pressure has also been observed under indentation testing[26] and diamond anvil cell (DAC) compression[14]. Amorphous B$_4$C may represent a reversion product from a shock-induced high-pressure phase occurring during decompression[14].

DAC hydrostatic compression of single crystal B$_4$C reveals uniform compression of B$_4$C to 60 GPa with no evidence of amorphous phase[14]. The same test conducted with hard granular pressure transmitting medium (PTM) leads to extensive amorphization of boron carbide. Amorphization is reported to be a consequence of nonhydrostatic stress and deformation brought about by the granular PTM[14]. Amorphous B₄C is observed following compression to peak pressures in excess of 25 GPa and subsequent decompression to below 20 GPa. Observations of amorphous B₄C are supported by both Raman spectra and post-test high-resolution electron microscopy. Amorphous B₄C is observed in heterogeneously distributed nanometer-thickness deformation lamella or shear bands. The extent of amorphization increases with pressure amplitude above the 25 GPa initiation pressure.

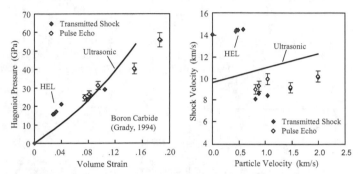

Figure 6. Hugoniot states from structured shock wave measurements on boron carbide ceramic using both the transmitted shock and pulse-echo analysis methods[8,12]. Comparison is made with extrapolation of ultrasonic compressibility[21].

Supporting molecular dynamics (MD) calculations of compression of a boron carbide unit cell consisting of a rhombohedra array of B₁₁C icosahedra with cross-linking C-B-C chain suggests sensible molecular mechanisms leading to observed pressure- and deformation-induced amorphization[14]. Results of the MD calculations are shown in Fig. 7. Hydrostatic compression of the unit cell to 60 GPa leads to self-similar compression with no evidence of crystallographic phase transformation. In contrast, nonhydrostatic uniaxial compression along the cross-link C-B-C chain leads to deformation instability at approximately 19 GPa uniaxial pressure through flexing of the cross-link chain. An approximately 4% volume reduction occurs between about 19 GPa and 22 GPa. Chain flexing continues to 44 GPa at which point permanent covalent bonding between cross-link boron and icosahedra boron occurs. Subsequent compression indicates mode softening and accompanying reduction in elastic compressibility. Amorphization is explained by the energetic state of the retained flexed cross-link chain and distorted rhombohedra unit leading to disruption of crystal stability on decompression. Both the hydrostatic and uniaxial compression resulting from the MD analysis are shown in Fig. 7. Comparison with the Birch EOS extrapolation of ultrasonic measurements on polycrystalline boron carbide[21] is also provided. Overlay with the McQueen Hugoniot measurements for boron carbide is shown on the right.

Both DAC experiments and MD simulations support the necessity of non-hydrostatic stress and deformation to initiate phase transformation and amorphization in boron carbide. Non-hydrostatic stress and deformation is intrinsic to the shock compression event. Shock compaction of porous boron carbide of McQueen is accompanied by intense microstructure stress heterogeneity and localized shear deformation. Similar, but less intense, deviator stress and shear deformation is incurred in the structured-wave shock compaction of full-density boron carbide ceramic[8].

The DAC tests and MD calculations of boron carbide exhibit encouraging consistencies with the inference of shock-induced phase transformation from the shock-compression Hugoniot data of McQueen[11,12]. Shock transformation at Hugoniot pressure at some point below 27 GPa, with substantial transformation volume, is in line with the DAC experiments indicating amorphization in the 20 to 25 GPa range and with MD observations of approximately 4% volume reduction within the 19-22 GPa onset of B_4C cross-link instability. Similarly, MD observations of atomic bonding transition and abrupt change in bulk compressibility at approximately 44 GPa is consistent with the observed shock transition of 45-50 GPa shock pressure with second-order transformation characteristics inferred from the Hugoniot data of McQueen.

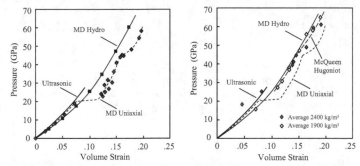

Figure 7. Molecular dynamics simulations[14] for hydro and uniaxial loading of a boron carbide crystal unit is shown on the left and compared with ultrasonic extrapolation[21]. Molecular dynamics simulations are compared with the McQueen Hugoniot for boron carbide on the right.

Although intriguing similarities between the McQueen Hugoniot data and the MD calculations are observed in Fig. 7, a rather significant discrepancy in compliance of the boron carbide is observed. First, the bulk compressibility assessed from the ultrasonic study of Manghnani et al.[21] is notably stiffer than the MD simulation of hydro compressibility. As noted previously, this difference is qualitatively consistent with the constraint imposed by the polycrystalline character of the ultrasonic test samples. Whether a quantitative assessment would account for the difference is not known. Second, the MD uniaxial calculations showing structural transitions in the B_4C crystal is markedly more compliant than the McQueen data. As will be subsequently shown, the McQueen Hugoniot data is the most compliant of available Hugoniot data for boron carbide. This discrepancy speaks either to difficulties with the intermolecular potential model use in the MD simulations, or to the possibility that only some fraction of the shocked material in the McQueen tests undergoes the suspected transformation. Note that the McQueen data is in reasonable good accord with the hydro MD calculation.

FURTHER SHOCK HUGONIOT DATA FOR BORON CARBIDE

The present study has primarily focused on analysis of the unique Hugoniot data of McQueen for boron carbide provided in the Los Alamos Shock Compendium[10]. Although these data were acquired on sintered ceramics with porosities ranging over 3% to 25%, the analysis shows these data to be internally consistent and to demonstrate some unusual compressibility features. Ample additional shock Hugoniot data for boron carbide are available and a systematic comparison with that of McQueen is warranted. These latter data include the studies of Wilkins[15], Gust and Royce[16], Pavlovski[17], Vogler *et al.*[4] and Zhang *et al.*[5]. In contrast to the work of McQueen, these other studies were performed on near-full-density boron carbide ceramic. The authors all report densities within 1% of theoretical. Comparisons of the McQueen data with the several higher-pressure Hugoniot points of Grady[8] for boron carbide were described earlier in the report.

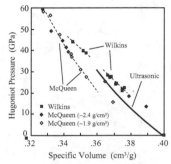 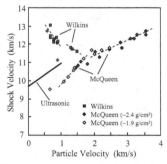

Figure 8. Shock Hugoniot data of Wilkins[15] on full density boron carbide. Comparisons with Hugoniot data of McQueen.

The Hugoniot data of Wilkins[15] are plotted with that of McQueen in Fig. 8. On the left are the shock pressure versus specific volume states. On the right is a comparative shock velocity versus particle velocity plot of the same data. Again, this plot is the normalized U vs. u representation of the McQueen data as described earlier. Correspondingly, that of Wilkins is also a normalized U vs. u plot of the data. Namely, the shock velocity, and the particle velocity, relate to the chord connecting the initial solid (theoretical) specific volume for the boron carbide ceramic to the final Hugoniot state through Eqns. (2). The double shock accessing the intermediate Hugoniot elastic limit state in the experimental shock velocity versus particle velocity analysis of the data is ignored in this representation. When used with care this description of the Hugoniot data can be informative as has already been illustrated in the earlier analysis of the McQueen data.

The lower pressure cluster of Wilkins' data points between 20-30 GPa are reasonably consistent with the three lowest pressure points (solid diamonds) of the McQueen higher density material. The samples of McQueen for the three tests were approximately 5% porous whereas those of Wilkins were close to full density (reported density was 2500 kg/m³). Both sets of data in this range indicate some residual strength when compared with extrapolated hydrostatic compressibility determined from the ultrasonic study of Manghnani et al.[21]. Of more interest is comparison of the Hugoniot points in the neighborhood of 40 GPa. The three higher density (solid diamonds) points of McQueen correspond to material with about 6% porosity. Normalized shock velocity plotted on the right show an increase with increasing shock amplitude consistent with a positive pressure derivative of the modulus in this range (~1.6 km/s particle velocity). Comparable data of Wilkins near 40 GPa show a decreasing shock velocity, suggesting reduction in retained strength or other unknown influences on the compressibility in this pressure range. At about 40 GPa Hugoniot pressure the difference in volume strain between the McQueen and Wilkins data is nearly 20%. Both plots of the data trend toward convergence at about 50 GPa. Compression of the high porosity material (22-25% porosity) of McQueen is dramatically at odds with the Wilkins data over the range of 15-40 GPa.

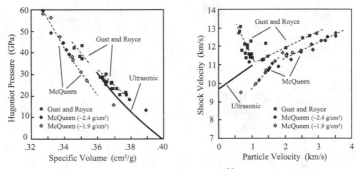

Figure 9. Shock Hugoniot data of Gust and Royce[16] on full density boron carbide. Comparisons with Hugoniot data of McQueen.

The same plots of the Hugoniot data for boron carbide ceramic provided in the study of Gust and Royce[16] are shown in Fig. 9. All data below 60 GPa were obtained with the inclined mirror technique. Four Hugoniot states in excess of 60 GPa shown in the normalized U vs. u plot used the flash gap method. Density of the Gust and Royce material was reported to be 2500 kg/m³. Hugoniot data analysis of Gust and Royce applied both the impedance match method and the free surface approximation (Hugoniot particle velocity equal to twice the measured free surface velocity). The data shown are from the latter method. Hugoniot results from the two analysis techniques were not statistically significant. Trends of the data are similar to that of the Wilkins data when compared with the Hugoniot curve of McQueen. The four higher pressure Hugoniot states (>60 GPa) are measurably stiffer (higher shock velocity) than comparable points of McQueen in the right hand U vs. u plot.

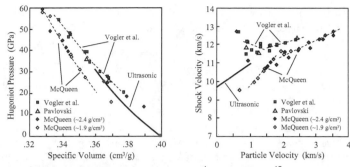

Figure 10. Shock Hugoniot data Vogler *et al.*[4] and Pavlovski[17] on full density boron carbide. Comparisons with Hugoniot data of McQueen.

Plots of Hugoniot data provided from the study of Vogler *et al.*[4] are compared with the McQueen data in Fig. 10. Average density of this ceramic obtained from Cercom Inc. was 2508 kg/m³. Data were obtained using the VISAR technique[25] to monitor time-resolved structured shock profiles. Again, Hugoniot states in the neighborhood of the three lower pressure points (15 to 25 GPa) of McQueen on the higher density (5% porous) ceramic are reasonably consistent. Marked differences between the two studies are observed at higher pressures. The shock velocity in the normalized U vs. u plot of the Vogler *et al.* data is fairly flat at about 12 km/s over the particle velocity range of approximately 1 to 2 km/s (Hugoniot pressures of about 25 to 55 GPa). This trend is unusual if the data describe reasonably normal lattice compression on the Hugoniot over this range and if material strength is not affecting the data. Compressibility does increase at the upper range of the Vogler *et al.* data, and Hugoniot states reside consistently above the McQueen Hugoniot.

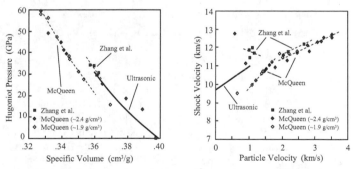

Figure 11. Shock Hugoniot data Zhang *et al.*[5] on full density boron carbide. Comparisons with Hugoniot data of McQueen.

Hugoniot data of Zhang *et al.*[5] using inclined mirror techniques are shown with the McQueen data in Fig. 11. Reported density for this boron carbide ceramic is

2516 ± 6 kg/m³. Although sparse, the data are reasonably consistent with the previous studies on full-density boron carbide. The cluster of Hugoniot points in the range of 25 to 35 GPa differ by about 25% in volume strain from the compression measured by McQueen. The Hugoniot states of McQueen in this range are on the high porosity material (open diamonds). The two higher pressure states of Zhang et al., on the other hand, are in reasonable agreement with the McQueen Hugoniot, as observed in the U vs. u plot on the right.

Pavlovski[17] reports four Hugoniot pressure measurements for a full density (2510 kg/m³) – one at an exceedingly high 220 GPa shock pressure. The relevant Hugoniot states of Pavlovski are plotted in Fig. 10 with the data of Vogler et al. The data are comparable in amplitude to that of Vogler et al. and suggest the same trend.

The various data on near-full-density boron carbide ceramic all show a tendency, to varying degrees, to exhibit markedly less compression than does the McQueen Hugoniot over the pressure range of about 15-40 GPa. In the lower portion of this range the observation refers to compressions achieved with the lower density materials (22-25% porosity) tested in the McQueen work. This observation ignores the several lower pressure points of McQueen on the higher density ceramic that exhibit clear strength effects.

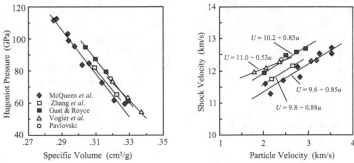

Figure 12. Comparisons of boron carbide Hugoniot data at the upper pressure range.

The data of the other authors on near-full-density ceramic also suggest a tendency to trend toward the McQueen data at higher pressure. Comparative plots of the Hugoniot data are provided in Fig. 12 to better display the higher pressure behavior over approximately 50-110 GPa. These plots of both the Hugoniot pressure vs. specific volume and the normalized shock velocity vs. particle velocity show that considerable differences among the data remain in this higher pressure range. The Hugoniot states of Gust and Royce, Pavlovski, and Vogler et al. all reside above that of McQueen. The several tests of Zhang et al. are reasonable close to the McQueen states. At a representative Hugoniot pressure of 80 GPa the spread in volume strain is about 8%. This difference is significant and is well outside the experimental scatter of any one experimental data set. Where reported this difference is also substantially larger than the experimental uncertainty. Also, for comparative purposes, the individual sets of data within this pressure range were fit to a linear $U = C_o + Su$ relation in the normalized

U vs. *u* plot. Values are 0.85, 0.88, 0.85 and 0.53 for *S* and 9.64, 9.80, 10.22 and 11.00 km/s for C_o, for the Hugoniot data of McQueen, Zhang *et al*, Gust and Royce, and Vogler *et al.*, respectively.

DISCUSSSION OF DISPARITIES AMONG HUGONIOT EOS DATA FOR B₄C

Summary

The present report has brought to the front the oft ignored Hugoniot data of McQueen *et al.*[9] (see Marsh[10]) for boron carbide ceramic. The data of McQueen are frequently dismissed in the high-pressure EOS assessment of boron carbide because of the significant and varied initial porosity of the test materials used in the study. Analysis of the data in this report accounts for the different initial porosities. The Hugoniot data are demonstrated to be internally consistent when porosity is accounted for. The experimental shock velocity versus particle velocity data of McQueen, which show wide variations due to the effects of porosity (3-7% porosity for the higher density set of data and 22-25% porosity for the lower density set), are mapped into a normalized representation of the shock velocity versus particle velocity EOS shock response of the material through the Hugoniot mass and momentum conservation relations. Plots of both this normalized display of the shock velocity versus particle velocity and the shock pressure versus specific volume reveal unique compressibility features for boron carbide on the Hugoniot.

A baseline equation-of-state compressibility for ambient pressure boron carbide is provided by the ultrasonic measurements of the elastic properties to 2.1 GPa[21]. Extrapolation of these elastic properties obtained on a full-density, high-quality polycrystalline ceramic with a Birch Murnaghan EOS relation provides a reasonable estimated of the compressibility to perhaps one order of magnitude higher pressure assuming lattice deformation remains self-similar. The ultrasonic EOS is in sensible agreement with subsequent ultrasonic measurements of Dodd *et al.*[22].

In the early 1990's time-resolved shock-wave profile measurements using velocity interferometry techniques were made on a full-density boron carbide ceramic[8]. Shock wave velocity histories reveal a unique elastic precursor structure hinting at complex failure mechanisms in this material. Stress relaxation and chaotic precursor wave structure suggest a time-dependent and heterogeneous nature to the dynamic shear failure of boron carbide ceramic. Later velocity interferometry measurements of Vogler *et al.*[4] on a boron carbide ceramic provided from another supplier show similar precursor structure in certain cases but not in others. In selected cases stable precursor structure comparable to similar measurements on other ceramics (*e.g.*, silicon carbide and aluminum oxide) are observed. Whether different material compositions or different test conditions are responsible for the marked precursor disparity was not determined. Several Hugoniot equation-of-state measurements from the study of Grady were in reasonable agreement with the shock compressibility measurements of McQueen. The Hugoniot measurements of Vogler *et al.* were not.

Hugoniot state measurements of McQueen range over shock pressures from about 15 to 115 GPa. Much of the lower pressure range is populated by tests performed on the lower-density (22-25% porosity) material whereas higher pressure states are achieved with the higher density (3-7% porosity) material. Appreciable overlap of the two data sets is achieved, however. Three lower pressure points on the higher density ceramic (~13 to 25 GPa) show evidence of strength and dynamic failure in the shock compression process. Compression states achieved in the more porous material within the pressure range of about 15 to 45 GPa are unexpectedly high when compared with the extrapolated hydrostatic compressibility from ultrasonic measurements. (Volume strains, from the initial solid state, are approximately 15-20% larger than the corresponding extrapolated hydrostatic compression.) It is unlikely that the ultrasonic extrapolation of self-similar elastic compressibility is in error to this magnitude for boron carbide, and suggests a possible change in lattice configuration (the I-II phase transition in Table 1) perhaps as low as the lowest ~16 GPa Hugoniot point of McQueen on the lower-density ceramic. Discounting this one point, the high porosity Hugoniot states of McQueen are consistently offset from the ultrasonic hydrostat from about 27 GPa shock pressure and higher. The sparse Hugoniot data of Grady tentatively suggests a compression anomaly between about 30 and 40 GPa shock pressure.

The collective higher pressure (approximately 25 GPa and higher) Hugoniot data of McQueen, when inspected in the normalized U vs. u plot, is not linear. Without too much imagination these data can be described by a bi-linear fit with intercept of the linear curves occurring in the neighborhood of 45 to 50 GPa. Such discontinuous behavior in the shock velocity versus particle velocity plane suggests a second phase transformation on the Hugoniot (the II-III transition identified in Table 1).

The recent diamond anvil cell (DAC) experiments and molecular dynamics (MD) analysis on boron carbide of Yan et al.[14] contribute significantly to the thinking on possible mechanisms affecting the dynamic equation-of-state response in the shock environment. DAC compression of single crystal B_4C reveals regions of material amorphization under pressure loading to in excess of 25 GPa, but only when subjected to a hard granular pressure transmitting media inducing severe heterogeneous stress concentration. Amorphous regions appear as thin lamella exhibiting features of inelastic shear deformation. MD simulations identify onset of non self-similar lattice compressibility relating to instability and distortion of the C-B-C cross-link chain at load pressures of about 20 GPa. Further, at about 45 GPa, MD calculations suggest an abrupt change in bulk elasticity associated with bonding of cross-link chain and icosahedra boron. The observed MD analysis phenomena are dependent, however, on the application of non-hydrostatic stress loading to the boron carbide crystal lattice. The same MD simulation applying strictly hydrostatic compression leads to self-similar compression of the B_4C to at least 60 GPa. These static DAC high pressure experiments and physics-based MD analysis of the response of boron carbide are shown to exhibit intriguing commonalities with the shock EOS of McQueen.

Difficulties emerge, however, in interpretation of the shock EOS of boron carbide when comparisons of the McQueen data are made with concurrent and later shock Hugoniot data of Wilkins, Pavlovski, Gust and Royce, Vogler et al. and Zhang et al. acquired on full or near-full density boron carbide ceramic. Stark differences in shock compression and other features of the shock data are observed, raising serious questions concerning the underlying causes of these unsettling disparities.

Discussion of Disparities

Experimental error is certainly a consideration in addressing the observed differences among the reported shock Hugoniot data for boron carbide. To at least some level, experimental differences have to be a factor. Test methods and measurement diagnostics have evolved markedly over the almost 40 years over which Hugoniot data for boron carbide have been collected. Whether experimental uncertainty has improved is another matter. Where reported experimental errors within a few percent are indicated. Systematic errors due to unknown experimental issues are always a possible factor.

The Hugoniot data of Gust and Royce[16] and Vogler *et al.*[4] are reasonably consistent with each other as are the several points of Pavlovsky[17] over much of the range of the data. These data are notably stiffer than the Hugoniot of McQueen. The several high pressure points of Zhang *et al.*[5] are close to the McQueen data, as are the two points of Grady in the mid pressure range. All other data, with the exception of that of Grady, are markedly stiffer than the McQueen Hugoniot in the intermediate 30-50 GPa shock pressure range.

Thus, in general the Hugoniot compression states of McQueen for boron carbide generated on initially porous ceramics would appear to be at odds with most other Hugoniot data. An argument for sensible accuracy of the McQueen Hugoniot for boron carbide is that the same experimental method and data analysis were applied to the measurement of shock-wave properties for a vast number of other solid materials – including other ceramics. These shock Hugoniot data, compiled in the Los Alamos Shock Wave Compendium[10] have generally compared well with later studies. A similar argument can be offered for the several Hugoniot measurements of Grady on boron carbide[8]. Namely, the same time-resolved VISAR diagnostic was applied to other light-armor ceramics in the same time period, including aluminum oxide, silicon carbide, tungsten carbide and titanium diboride. Some experimental differences among the various Hugoniot data cannot be ruled out. But, other explanations for the substantial disparities should be explored.

One explanation that has been considered previously by Dandekar[3] is that Hugoniot states measured on full-density boron carbide ceramic exhibit a material strength offset. Compression states achieved on the more porous ceramics of McQueen potentially reside closer to the hydrodynamic EOS response of the material because of the more extensive deformation dissipation and temperature rise incurred in the shock compression process. This strength offset consideration applies to the more extensive data sets of Wilkins, Gust and Royce, and Vogler *et al.*, and ignores the several Hugoniot points of Grady which do not support this argument. Based on an assumption of strength offset Dandekar provides a reasonable model for the dynamic strength of boron carbide ceramic under shock compression.

The difficulty with an explanation based on strength offset for the discrepancy among the various Hugoniot data is provided by the study of Vogler *et al.*[4]. Experiments were performed using VISAR diagnostics where the material is first shocked to the Hugoniot state and subsequently (within less than one microsecond) subjected to either a second shock wave that loads the material to an even higher shock pressure state, or to a release wave that decompresses the material from the Hugoniot state. Through analysis of these secondary waves an assessment of the deviator stress (stress offset) at the

Hugoniot state is determined. The analysis of Dandekar discussed previously assumes that the stress offset and the strength of the material at the Hugoniot state are one and the same. This assumption is not accepted in the investigation of Vogler *et al.*, and the measurement is designed to independently determine the stress offset and the strength of the material at the Hugoniot state. The second shock is particularly informative in this regard. If, for example, the stress offset and strength are comparable then the first shock that carries the material to the Hugoniot state will be followed by a second single shock wave traveling at the nominal bulk wave speed of the material. This behavior is schematically illustrated in Fig. 13 where the Hugoniot state with stress offset is identified by the number 1 on the right. The initial and following shock wave should look as illustrated by the corresponding stress-time history on the left. If, on the other hand, a Hugoniot state is achieved with negligible stress offset (point 2 in the right hand figure) then the second shock will be a structured wave with both an elastic precursor wave and a following shock wave. The second stress history on the left identifies this case. The first example is representative of Hugoniot strength assumption pursued by Dandekar for boron carbide[3]. This approach is justified by previous studies on other materials.

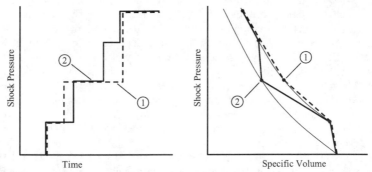

Figure 13. An elastic-plastic material illustrating shock pressure paths and wave profiles for a principal Hugoniot state exhibiting both negligible and significant stress offset.

The extensive study of Vogler *et al.*[4], however, shows that the shock response of boron carbide corresponds more closely to the latter example. Namely, that the principal Hugoniot states in excess of about 30 GPa lie close to the hydro or EOS compression of the material with quite small stress offset. Such a transient loss in strength has previously been suggested[8], and is in line with the proposition of Chen *et al.*[13] that such strength loss results in reduce ballistic performance of boron carbide ceramic. Both a subsequent shock compression wave, and a release decompression wave, reveal material strength through structured waves with elastic and plastic (inelastic) components. Implications of the Vogler *et al.* study are that discrepancies between the McQueen Hugoniot data for boron carbide and the several other sets of Hugoniot data are not explained by a stress offset. This aspect of the study of Vogler *et al.* on boron carbide should be, however, viewed critically. Wave interactions are complex in this technique and analysis of the data requires several assumptions that, while reasonable, add to the uncertainty. Perhaps the strongest support for the conclusions arrived at for boron carbide in the Vogler et al.[4] investigation is provided by a similar study on silicon carbide ceramic (Vogler et al.[27]). Similar testing and analysis shows that silicon carbide, in contrast, exhibits significant stress offset at the Hugoniot state.

A further explanation for the disparity can be entertained. The diamond anvil cell (DAC) amorphization experiments and accompanying molecular dynamics (MD) analysis suggest a mechanism for the observed differences between the McQueen Hugoniot data and the markedly less compressible Hugoniot measurements of Wilkins[15], Gust and Royce[16], and Vogler et al.[4]. DAC experiments of Yan et al.[14] show evidence of amorphization of boron carbide single crystal under high pressure. The authors suggest amorphization as a recovery state resulting from a pressure (and stress) induced crystal structure transition upon decompression from elevated pressure. The occurrence of such a structure transition is dependent on the application of distortional stress and deformation during the pressurization process. Yan et al. report amorphization entrained within laminar shear bands exhibiting shear displacement through deformation plasticity. Further, first principles MD analysis suggest a crystal structure transition through onset of bending instability of the cross-link C-B-C chain under nonhydrostatic stress loading in excess of approximately 20 GPa. The structure transition is accompanied by an approximately 4% volume change.

An important difference is noted in the shock compression of the near-full-density boron carbide ceramics tested by Wilkins, Gust and Royce, and Vogler et al, and the significantly more porous material of McQueen. In all cases a degree of deformation plasticity is necessary to accommodate the uniaxial strain constraint of the shock compression event. Collapse of void volume of a porous solid subjected to shock compression, however, involves markedly more intense inelastic deformation and dissipation.

The study of Yan et al. suggests that amorphization, and presumably crystal structure transformation, is to some extent proportional to the degree of distortion stress and deformation impressed on the material. If this is the case then it is reasonable to suppose that such a transformation with appreciable transformation volume change is occurring under shock compression in all of the shock studies conducted. Transformation progression, however, would be substantially further along at comparable shock pressures in the McQueen material due to the more extensive shear deformation brought about by pore compaction during the shock compression process. The close proximity of Hugoniot states for both the low and the high porosity ceramics of McQueen at pressures in the neighborhood of 40 GPa and higher suggests the transformation is near completion; subsuming nearly 100% of the material. The flat or decreasing trend of the shock compressibility of the Wilkins, Gust and Royce, and Vogler et al. data in the 35 to 45 GPa shock pressure range suggests similar transformation may also be proceeding in the shock compression of these higher density ceramics. Whether the same extent-of-transition argument can be extended to explain the nearly 8% volume strain discrepancy in the 60 to 80 GPa shock pressure range is arguable. The following thought can be offered, however. MD calculations of Yan et al. suggest that no transformation proceeds under purely hydrostatic loading to 60 GPa. One can speculate that heterogeneous deformation in the shock compression process could subject some fraction of the material (isolated regions between deformation shear bands) to near hydro loading to quite high pressures, retaining some portion of the material in the lower pressure phase.

Independent analysis of the McQueen Hugoniot data earlier in the paper identified two Hugoniot phase transformations (anomalies in the high pressure shock Hugoniot). The first I-II transition occurs (or initiates) below 27 GPa. Comparison with the initial phase compressibility assessed from ultrasonic measurements suggests appreciable transformation volume strain. This transformation may relate to observed amorphization and the shock compression discrepancy discussed in the earlier paragraphs.

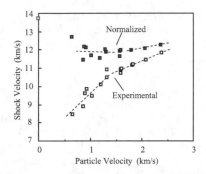

Figure 14. Experimental and normalized shock velocity versus particle velocity representation of the Vogler et al.[4] boron carbide shock Hugoniot data.

The second II-III transition is identified from a discontinuity in the slope of the shock compressibility data at about 45 to 50 GPa. Second order characteristics with little or no transition volume are associated with this transition. In the crystal structure transition model proposed by Yan *et al.*, this transformation may relate to the irreversible bonding of boron between the cross-link chain and icosahedra crystal components. A comparable discontinuity in slope of the experimental shock velocity versus particle velocity data is reported at shock pressures of 38 to 40 GPa in the studies of Vogler *et al.* and Zhang *et al.* This break in slope is not apparent, however, in the normalized U vs. u plot of the data (see Figs. 10 and 11). The experimental and normalized U vs. u plots of the boron carbide Hugoniot data of Vogler *et al.* are compared in Fig. 14 illustrating the differences and providing an interesting perspective of the shock data. The observed discontinuity reported by the authors may relate to a transition in strength on the Hugoniot. See for example figure 12 in the study of Vogler *et al.* where the transition to minimal stress offset in the strength behavior on the Hugoniot is displayed.

As reasonable as it may be, one problem with the explanation offered for the differences between the McQueen and other Hugoniot data is the several Hugoniot measurements of Grady[8] on a full-density and high-quality boron carbide ceramic that are in good agreement with the McQueen data in the critical 40 to 50 GPa shock pressure range. As difficult as it is for a scientist to have to reject his own data this may have to be the case. No reasonable explanation for the stark differences with the other Hugoniot data on full or near-full density boron carbide comes to mind. Experimental error is certainly a possibility. As noted previously, however, the same experimental method in the same time frame was used to test other ceramics and these data have held up in comparisons with later work. It is also worth noting that the Hugoniot data of Gust and Royce[15] and of Vogler et al.[4] on near-full-density boron carbide ceramic are in sensible agreement with both the McQueen and the Grady data up to Hugoniot pressures of about 30 GPa – the region where material strength plays an important role.

It has also been pointed out to the author (Palicka[28]) that stark differences in materials and material preparations could also be responsible for the observed differences. In the 1960's there were two primary suppliers of armor-grade boron carbide ceramic in the U.S. – Norton Co. and Carborundum Co. The former excelled in hot pressure-assisted preparation whereas the latter relied principally on pressureless sintering methods. The materials of Wilkins[15] and Gust and Royce[16] were provided by Norton Co. The source for the material of McQueen et al.[9,10] is not known.

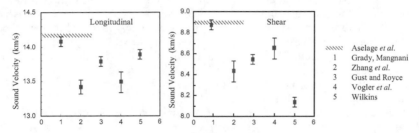

Figure 15. Longitudinal and shear sound velocities reported for test materials used in shock Hugoniot experiments. Baseline sound velocity is provided by the study of Aselage et al.[2].

Although not an explanation for the disparity, one intriguing difference with other shock studies should be noted. In the initial characterization of materials prior to shock wave testing all authors provided measurements of density and ultrasonic wave speeds of the test samples. Density of the boron carbide tested by Grady is within 1% of densities determined by other workers and with the theoretical density of boron carbide. In contrast, marked differences in ultrasonic sound speeds are reported. Sound velocity measurements from the various studies are shown in Fig. 15. The upper bound dash towards the top of the graph can be reasonable accepted as the correct longitudinal and shear wave speeds for full density polycrystalline boron carbide. These data reported in Aselage et al.[2] and also in Gieski[29] were performed on research grade polycrystalline samples prepared at Sandia National Laboratories, and are consistent with other comparable studies referenced in that paper. Sound speeds in the ceramics studied by Grady[8] and also by Manghnani et al.[21] provided by Dow Chemical Co. are consistent with the Aselage measurements. The marked reduction in sound speeds in other test materials is not explained. Reduced carbon in B$_4$C will lead to low sound speeds[2], although the extent of carbon reduction required is considerable and thus this is unlikely to be the cause. Aselage et al.[2] note that microcracking in boron carbide ceramic can occur in the preparation process of polycrystalline ceramic and will lead to reduced sound speed measurements. A micro crack density may well be the explanation for reduced sound speeds in the other studies. It is, however, difficult to see how microstructure cracking could appreciably influence the shock equation of state.

CLOSURE

In summary, the consensus of the present survey of the available shock Hugoniot data for boron carbide is that compression of the B_4C crystal structure is not self similar within the ballistic shock pressure range of 0-60 GPa. Rather, boron carbide undergoes two possible polymorphic phase transformations induced by the shock compression process. A best estimate of the transformation pressures and volumes along with hydrodynamic compressibility properties of the ambient phase and the two high-pressure phases is provided in Table 1. The inferred transformations gain some support from the recent DAC experiments and MD analysis on boron carbide of Yan *et al.* A crucial element from these latter studies is the potential sensitivity of transformation to shear deformation. This feature points to a possible explanation for differences observed between the McQueen shock data on more porous ceramic, where substantially more shear deformation and dissipation is incurred in the shock process, and the preponderance of data from other workers on full density boron carbide. Namely, shear dissipation transformation is occurring in all of the data considered, however the extent of transformation at the same pressure is dependent on the degree of shear dissipation incurred in shock compression to the Hugoniot state.

Dynamic strength and crystallographic phase transition of boron carbide in the shock and ballistic environment appear inextricably mixed. Transient loss of strength and subsequent strength recovery is demonstrated by structured shock wave studies of later workers. Such results are in line with observations of reduced ballistic performance of boron carbide in the post-failure regime.

The conclusions reached in the present study are certainly arguable, but do represent a reasonable assessment of the available data. Boron carbide is and will probably remain one of the most enigmatic ceramics under investigation by the shock and ballistic community.

REFERENCES

[1] D. Emin, Structure and Single Phase Regime of Boron Carbide, *Phys. Rev. B*, 38, 6041 (1988).

[2] T. L. Aselage, D. R. Tallant, J. H., Gieske, S. B. VanDeusen, R. G. Tissot, Preparation and properties of icosahedral borides, in *The Physics and Chemistry of Carbides, Nitrides, and Borides*, R. Freer, ed., NATO ASI Series E, Vol. 185, pp. 97-111, Kluwer Academic Publishers, Dordrecht (1990).

[3] D. P. Dandekar, Shock Response of Boron Carbide, *Army Research Laboratory Report* ARL-TR-2456, April (2001).

[4] T. J. Vogler, W. D. Reinhart, L. C. Chhabildas, Dynamic Behavior of Boron Carbide, *J. Appl. Phys.*, 95, 8, 4173 (2004).

[5] Y. Zhang, T. Mashimo, Y. Uemura, M. Uchino, M. Kodama, K. Shibata, K. Fukuoka, M. Kikuchi, T. Kobayashi, T. Sekine, Shock Compression Behaviors of Boron

Carbide, *J. Appl. Phys.* 100, 113536 (2006).

[6] T. J. Holmquist and G. R. Johnson, Characterization and Evaluation of Boron Carbide from One-Dimension Plate Impact, , *J. Appl. Phys.*,100, 093525 (2006).

[7] J. A. Ciezak and D. P. Dandekar, Compression and Associated Properties of Boron Carbide, in *Shock Compression of Condensed Matter*, M. L. Elert et al., eds, AIP Press, 1287-1290 (2009).

[8] D. E. Grady, Shock Wave Strength Properties of Boron Carbide and Silicon Carbide, *International Conference on Mechanical and Physical Behavior of Materials under Dynamic Loading*. les Editions de Physique, pp. 385-391 (1994).

[9] R. G. McQueen, S. P. Marsh, J. W. Taylor, J. N. Fritz, W. J. Carter, The Equation of State of Solids from Shock Wave Studies, In *High Velocity Impact Phenomena*, R Kinslow ed., Academic Press (1970).

[10] S. P. Marsh, LASL Shock Hugoniot Data, Edited by SP Marsh, University of California Press, Berkeley (1980).

[11] D. E. Grady and C. Doolittle, Preliminary Compressibility Analysis of Boron Carbide from High-Pressure Shock and Ultrasonic Data, *Applied Research Associates Rept.*, for U.S. Army TACOM-TARDEC, February (2001).

[12] D. E. Grady, Analysis of Shock and High-Rate-Date for Ceramics: Application to Boron Carbide and Silicon Carbide, *Applied Research Associates Rept.*, for U.S. Army TACOM-TARDEC, August (2002).

[13] M. W. Chen, J. W. McCauley, K. J. Hemker, Shock Induced Localized Amorphization in Boron Carbide, *Science*, 299, 1563 (2003).

[14] X. Q. Yan, Z. Tang, L. Zhang, J. J. Guo, C. Q. Jin, Y. Zhang, T. Goto, J.W. McCauley, M. W. Chen, Depressurization Amorphization of Single-Crystal Boron Carbide, *Phys. Rev. Letts.*, 102, 075505 (2009).

[15] M. L. Wilkins, Third Progress Report of Light Armor Program, Lawrence Radiation Laboratory, Livermore, CA, Report No. UCRL-50460 (1968).

[16] W. H. Gust and E. B. Royce, Dynamic Yield Strength of B_4C, BeO and Al_2O_3 Ceramics, *J. Appl. Phys.*, 42, 276 (1971).

[17] M. N. Pavlovskii, Shock Compressibility of Six Very Hard Substances, *Sov. Phys. Solid State* 12, 1737 (1970).

[18] T. J. Holmquist and T. J. Vogler, The Response of Silicon Carbide and Boron Carbide Subjected to Shock-Release-Reshock Plate-Impact Experiments, DYMAT 2009, 119-125, EDP Sciences, DOI: 10.1051/Dymat/2009016 (2009).

[19] T. J. Holmquist and G. R. Johnson, Response of Silicon Carbide to High Velocity Impact, *J. Appl. Phys.*, 91, 5858-5866 (2002).

[20] N. S. Brar, Z. Rosenberg, S. J. Bless, Applying Steinberg's Model to the Hugoniot Elastic Limit of Porous Boron Carbide, *Shock Waves in Condensed Matter - 1991*, Elsevier Science, pp. 451-454 (1992).

[21] M. H. Manghnani, Y. Wang, F. Zinin, W. Rafaniello, Elastic and Vibration Properties of B_4C to 21 GPa., *Science and Technology of High Pressure Volume 2*, Proc. Int. Conf. on High Pressure Science and Technology (AIRAPT-17), July 25–30, University Press, 945–948 (2000).

[22] S. P. Dodd, G. A. Saunders, B. James, Temperature and Pressure Dependences of the Elastic Properties of Boron Carbide, *J. Mater. Sci.*, 37, 2731-2736 (2002).

[23] F. Birch, Finite Strain Isotherm and Velocities in Single-Crystal and Polycrystalline NaCl at High Pressures and 300 K, *J. Geophys. Res.* 83, 1257–1266 (1978).

[24] R. Jeanloz, Shock Wave Equation of State and Finite Strain Theory, *J. Geophys. Res.*, 81, 5873 (1989).

[25] L. M. Barker and R. E. Hollenbach RE, A Laser Interferometer for Measuring High Velocities of any Reflecting Surface, *J.Appl. Phys.*, 43, 4669-4675 (1972).

[26] X. Q. Yan, W. J. Li, T. Goto, M. W. Chen, Raman Spectroscopy of Pressure-Induced Amorphous Boron Carbide, *Appl. Phys. Letts.*, 88, 131905 (2006).

[27] T. J. Vogler, W. D. Reinhart, L. C. Chhabildas, D. P. Dandekar, Hugoniot and Strength Behavior of Silicon Carbide, *J. Appl. Phys.*, 99, 023512 (2006).

[28] R. Palicka, private communication (2010).

[29] J. H. Geiske, T. L. Aselage, D. Emin, Elastic Properties of Boron Carbide, in Boron Rich Solilds, D. Emin *et al.,* eds., *AIP Conf. Proc.* N0. 231, AIP New York, p. 376.

MULTISCALE MODELING OF ARMOR CERAMICS

Reuben H. Kraft
U.S. Army Research Laboratory
High-Rate Mechanics and Failure Branch
Aberdeen Proving Ground, Maryland, United States

Iskander (Sasha) Batyrev
U.S. Army Research Laboratory
Energetic Materials Science Branch
Adelphi, Maryland, United States

Sukbin Lee
Departement of Materials Science and Engineering
Carnegie Mellon University
5000 Forbes Avenue
Pittsburgh, Pennsylvania, United States

A.D.(Tony) Rollett
Departement of Materials Science and Engineering
Carnegie Mellon University
5000 Forbes Avenue
Pittsburgh, Pennsylvania, United States

Betsy Rice
U.S. Army Research Laboratory
Energetic Materials Science Branch
Aberdeen Proving Ground, Maryland, United States

ABSTRACT

In this work a multiple length scale, multidisciplinary computational framework for simulating the response of ceramic materials subjected to dynamic loading is developed. The bridging between atomistic and mesoscopic length scales is addressed in a hierarchical fashion through the description of interfacial failure mechanisms and by passing bulk elastic properties. The mesoscopic length scale is treated using parallel, three-dimensional finite elements with microcracks explicitly represented on the grain boundaries using cohesive interface laws allowing investigation of crack nucleation, growth, and coalescence. The relationships at the atomic level are determined by molecular dynamics and quantum mechanical characterizations. While combining the individual pieces of the multiscale effort into a coherent framework are in the preliminary stages, this work demonstrates the feasibility of the approach and highlights areas where more attention is needed.

INTRODUCTION

The behavior of materials is controlled by their microstructure and the failure processes and mechanisms at various length and time scales. Therefore, multiscale modeling is necessary to understand the physics of failure and may also provide the ability to probe and investigate controlled

143

variations in material characteristics that yield effective means for microstructural design. Multiscale modeling approaches are commonly classified as hierarchical or concurrent. The primary ingredients of a computationally-based multiscale approach include, 1) a numerical framework capable of capturing the physics of each length and time scale of interest, and 2) a coupling method used to pass information between scales. For example, in hierarchical multiscale modeling, the coupling approach provides boundary conditions to small scale problems and gathers constitutive information to be used at larger scales.

In order to develop the computational framework for multiscale modeling of ceramics, we choose to model polycrystalline Aluminium Oxynitride (AlON) [16] because of the advantageous characteristics it has for model development, for example, it is transparent, apparently has a small of amount of grain boundary phase, has a cubic crystal structure, and has a large grain size compared to other ceramic materials in the range of 150-250 μm. Transparency can be helpful while trying to understand mechanisms of failure during experiments [20]. A large grain size will be helpful once mesoscale simulations are comparable to laboratory-sized specimens at the validation phase of the work.

At the atomistic length scale, in spite of multiple successful applications of oxinitrides still little known about its atomic structure. For example, where exactly is the Al vacancy, how are N atoms distributed along the O sites, or how far N atoms prefer to be from an Al vacancy. In this paper we try to shed some light on atomic structure and elastic properties of AlON from first principles calculations and then propose a means to use this information at larger length scales.

At the mesoscopic length scale, several attempts have been made to model ceramic materials, especially the inelastic response, which may include complex interactions of microcracks. Since the interaction of microcracks is believed to be important to the failure process, explicit representation of the cracks is a characteristic desirable of any model. Computational mesoscale modeling has been an interest to many scientists and engineers in the past, with considerable focus on brittle materials such as ceramic materials. Over time, advances in computational resources have allowed new avenues of simulation to emerge. For example, by introducing cohesive laws into the computational framework cracks can be modeled explicitly [1, 5, 18, 19, 26]. This is a powerful technique since crack interaction and coalescence can occur for complex geometries, such as through a microstructure, an advantage analytic-based approaches cannot offer. The introduction of cohesive elements lead to additional understanding of complex processes involved with dynamic fracture during impact events including fragmentation, spallation, and damage evolution [3, 4, 6, 7, 15, 18, 26]. Virtual microstructures were incorporated into the simulation framework, classified as mesoscopic, using Voronoi tessellation [6, 7, 17] and typically consist of at most a few hundred grains. Then, the introduction of an advanced contact algorithms into the mesoscale computational framework allowed for the simulation of two-dimensional microstructures subjected to multiaxial compressive loading [11, 12, 24]. In addition, both anisotropy and finite deformation plasticity have been implemented into grain-level finite element analysis to help understand deformation and failure responses in multi-phase polycrystalline materials [4].

METHODS

In this work, we address two length scales, the atomistic level and the mesoscale using two different numerical approaches. At the atomistic length scale the general gradient approximation (GGA) and projected augmented waves (PAWs) with energy cut-off of 600 eV were used for the simulations. We have assumed an "idea" stoichiometry of cubic AlON with 35.7 mole % AlN using the constant anion model, and assuming an Al vacancy (compensating for the excess of charge related with N atoms) on the octahedral site: $Al_8^{IV} Al_{15}^{VI} \square^{VI} O_{27} N_5$. The total energy of the system with vacancy at the octahedral site has ~ 0.5 eV lower energy than that at tetrahedral site (Figure 1a). The calculations have been carried out for two general structural models of nitrogen atomic arrangements in the unit cell: (I) random to clustered (Figure 1b) and (II) adjacent (Figure 1c) to not-adjacent (Figure 1a) to the Al octahedral vacancy.

At the mesoscale, the finite element method is used along with a multi-body contact algorithm and cohesive element technology to permit the inclusion of explicit microcracking into simulations. The basis of our model is the Lagrangian approach and finite element meshes representing computational microstructures are constructed using Voronoi tessellation (also known as Theissen or Dirichlet [2]) and a Monte Carlo-based grain growth approach. In both approaches, stereolithographic (STL) files which are surface representations of the grains are volumetrically meshed. This process of surface-to-volume meshing for computational microstructures is extremely important to successfully conduct mesoscale simulations and the impact of poor meshes will be discussed later. To create the surface meshes, equi-axed three-dimensional polycrystalline microstructures are generated using a standard Monte Carlo grain growth algorithm (see for example Rollett and Manohar [23]), which provides an image on a regular simple cubic grid where each point is associated with a grain or material number. The advantage of such an approach over, say, a Voronoi tessellation of a set of randomly positioned points is that the microstructure has more realistic distributions of grain size, nearest neighbor relationships and other microstructure measures. A multi-material marching cubes method derived from Wu and Sullivan [25] was used to interpolate a conformal triangular surface mesh into the image where each triangular element separates two different material numbers. The resulting mesh is, obviously, stair-stepped (aliased) and so a smoothing algorithm with constraints on the quadruple points (at junctions where sets of three triple lines meet) is applied to minimize curvature on the grain boundary facets. The method is based on the gradient-weighted moving finite element approach of Kuprat [13]. The resulting smoothed surface mesh in STL format then provides the basis for the volume mesh. The volume-meshed microstructures consist of grains which are treated as separate bodies in a multi-body contact/interface algorithm. Three sets of simulations were conducted, one for each of the meshes shown in Figure 2. Figure 2(a) shows a 50 grain mesh with 885,755 elements constructed using the Monte Carlo-based grain growth approach, Figure 2(b) shows a 1000 grain mesh with 2,563,510 elements constructed using Voronoi tessellation, and Figure 2(c) shows 1220 grains meshed using 2,128,337 elements constructed using the Monte Carlo-based grain growth approach. Currently, four-noded tetrahedral elements are being used.

The material properties for Aluminium Oxynitride (AlON) listed in Table 1 were used. At this point, only isotropic linear elasticity is being used for the bulk grain material. In the future, anisotropic linear elasticity will be used based on results of density functional theory calculations of AlON for elastic constants.

RESULTS & DISCUSSION

The development of a multiscale framework is challenging because adequate approaches for each length and time scale must be developed and verified separately then combined, therefore, in the early stages the work may seem disjoint. Here, results and discussions are given on both the atomistic and mesoscopic length scales that were conducted by different researchers. While the two different approaches have not be blended into a coherent package, the individual pieces show great promise and the precise mechanism of passing length scale information is more evident as a result of this work.

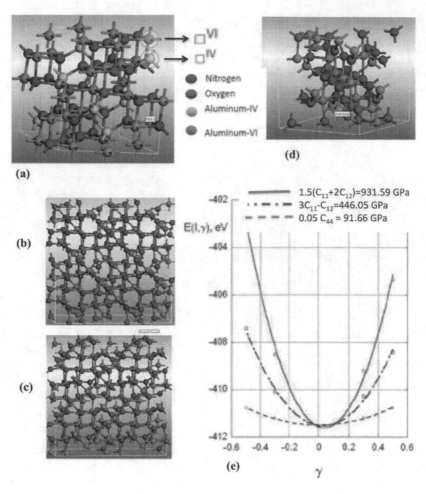

Figure 1: (a) Total energy of Al vacancy at octahedral and tetrahedral. Cluster (b) (average NN distance of N atoms d is 2.86 Å) versus random (c) (d=4.6 Å) distribution of N atoms along O sites. (d) Cluster distribution of N atoms around Al vacancy. (e) Parametrization of total energy for three independent strains used to calculate elastic constants of ALON with a uniform distribution of N atoms.

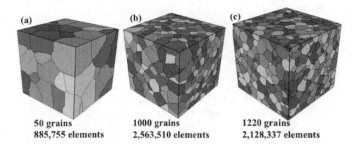

Figure 2: Three dimensional finite element meshes with a) 50 grains constructed using the Monte Carlo-based grain growth approach, b) 1000 grains developed using Voronoi tessellation, and c) 1220 grains using constructed using the Monte Carlo-based grain growth approach.

Table 1: Representative material properties for polycrystalline AlON.

Material Parameter	Value
Density, ρ	3673 kg/m^3, [20]
Young's Modulus, E	320 GPa, [16]
Poission's Ratio, ν	0.25, [16]
Shear Modulus	123 GPa, [16]
Flexure Strength, σ_c	307 MPa, [16]
Fracture Toughness, K_{IC}	2.9 MPa$\sqrt{\text{m}}$ [16]
Critical Energy Release Rate, G_c	26 J/m^2, Derived from K_{IC}

Atomistic Length Scale

A clustered distribution of N atoms away from the Al vacancy has $\sim$1 eV per 55 atoms higher total energy than for a random distribution. A clustered distribution of N atoms adjacent to the Al vacancy (Figure 1c) results in the same higher total energy of a system as that with a random distribution of N atoms bordering Al vacancy. It means that a cluster distribution is less energetically favorable than a random one, but if it exists in small concentrations, the cluster distribution is most likely around an Al vacancy. Independent strains were applied to a unit cell, parameterizing the total energy as a function of the strain to compute the elastic constants (Figure 1e). The purpose of the calculations is to determine if the location and/or segregation of N atoms in the unit cell affects the elastic properties of AlON. The type of distribution affects elastic constants, as seen from Table 2. The data is calculated for random and cluster distribution of N atoms not adjacent to the Al vacancy. The calculated elastic constants are in overall agreement with experimental measurements for polycrystalline ALON. The variation of C_{11}, C_{12} and C_{44} for random and cluster distributions of N atoms may be related with different degree of strain accumulated in a cubic shape unit cell with the cluster and random distributions of N atoms.

Table 2: Elastic constants calculated for random and cluster distribution of N atoms in comparison with experimental data for polycrystalline ALON.

Elastic Constant	Random	Cluster	Experiment
C_{11} (GPa)	306.15	365.65	369–393 [9, 14, 22]
C_{12} (GPa)	157.47	190.23	124–132 [9, 14, 22]
C_{44} (GPa)	183.33	153.72	123–131 [9, 14, 22]

Mesoscopic Length Scale

As previously mentioned, microcracks are modeled using cohesive interface laws. Typically cohesive laws assume that a relationship exists between traction and separation of two surfaces at an interface. The general idea is to impose a traction distribution over a certain length on the crack faces to remove the stress singularity. Physically, the distribution represents cohesion due to the breaking of atomic bonds ahead of the crack tip. This assumption immediately introduces additional length and time scales into the problem [21]. Numerically, the imposed stress distribution eliminates the stress singularity at the "real" crack tip by imposing tractions across a "virtual" crack tip, acting to close the crack, as schematically depicted in Figure 3a. The imposed distribution relates the crack opening displacement and tractions and is commonly referred to as a cohesive law. Herein, a linear-decreasing cohesive law is used, as also shown in Figure 3a. Using cohesive interfaces allow microcracks to propagate along interfaces (as seen in Figure 3b). A typical stress-strain response of a failure process between two grains is shown in Figure 4a.

One of the overarching assumptions made about modeling fracture in solids, is that the process is a toughness-controlled phenomenon. That is, fracture is controlled by K_I (or the fracture energy release rate, G_c). Thus, to simulate fracture processes as best possible, this concept of toughness-controlled must be verified and maintained when using the cohesive interface approach (or any other method). In numerical analysis, length scales such as mesh and specimen size can cause erroneous solutions if not selected properly. For example, Falk et al. [8] suggests the relationship

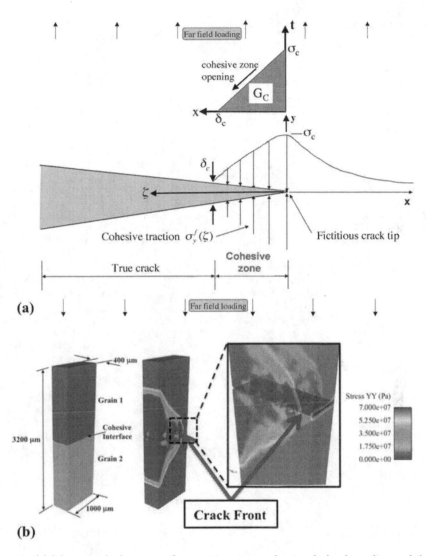

Figure 3: (a) Schematic of cohesive interface opening process, showing the local coordinates, failure stress distribution, "real" and "virtual" (cohesive zone) cracks, and far-field loading conditions. (b) Bi-crystal, one interface computational model with a pre-notch used to understand the cohesive fracture process associated with grain boundaries.

between mesh size l_m, cohesive zone length l_z, and the distance between cohesive interfaces as l_d:

$$l_d > l_z > l_m \tag{1}$$

where $l_d > l_z$ in order to eliminate problems that arise when cohesive surfaces exist between all elements that can be eliminated by using dynamic insertion of the cohesive interfaces. We point out an additional bound on the sample size in which a cohesive interface is embedded. The basic idea relates directly to the crack process zone. If the specimen's boundary edges are within the bounds of the process zone, fracture is no longer a toughness-controlled phenomenon, but instead is controlled by σ_c in the cohesive law, $i.e.$, simulations become strength-controlled. This effect comes as no surprise since the cohesive interface idea was founded upon a single crack in an infinite medium [1]. Kraft [10] found that the specimen size (measured from the crack interface to specimen boundary edge) should be at least 20 times the size of the cohesive zone length for the simulated fracture process to remain toughness-controlled. To demonstrate this point, a two-dimensional computational experiment is conducted where a specimen size at least 20 times the size of l_z is used and the critical fracture stress, σ_c in the cohesive law is changed while keeping G_c constant (thus we also change δ_c according to the cohesive law). If the process is toughness-controlled, changing σ_c should have no effect on the macroscopic fracture strength. The results for this computational experiment are shown in Figure 4b. The fact that the cohesive law parameters are changed, yet the macroscopic strength remains similar is a reassuring result. To summarize, when the sample size becomes comparable to the cohesive zone length scale the fracture process becomes strength-controlled. That is, the macroscopic fracture stress is governed by σ_c.

In order to obtain cohesive interface properties to describe the grain boundaries, recall the relationship given by elastic fracture mechanics between the mode-I stress intensity factor, K_I, and the energy release rate of the material, G_c given by $K_I^2 = E'G_c$ where $E' = E$ for plain stress or $E' = E/(1 - \nu^2)$ for plain strain. McCauley [16] reports that AlON has a fracture toughness of 2.4-2.9 MPa$\sqrt{\text{m}}$, therefore, in plane stress with $E = 320$ GPa and $K_{IC} = 2.9$ MPa$\sqrt{\text{m}}$, $G_c = 26$ J/m^2. In addition, since a linear cohesive interface law is used, where $G_c = \frac{1}{2}\sigma_c\delta_c$, the critical crack opening displacement, $\delta_c = \frac{2G_c}{\sigma_c} = 1.7 \times 10^{-7}$ m (assuming σ_c is equal to the flexure strength of AlON). This assumption may not be correct since the flexure strength is a bulk measured property from many grains, not just a single interface. It is important to note that bi-crystal micro-tension experiments would be extremely useful for determining grain boundary interface strength more accurately. The ramifications of using a critical fractures stress that is not correct can be illustrated by considering the cohesive zone length calculated from the material properties of AlON listed in Table 1:

$$l_z = \frac{9\pi}{32} \frac{E}{(1 - \nu^2)} \frac{G_c}{\sigma_c^2} = \frac{9\pi}{32} \frac{320\text{GPa}}{(1 - (0.25)^2)} \frac{26J/m^2}{(307\text{MPa})^2} = 83\mu\text{m} \tag{2}$$

An explanation and derivation for the equation for l_z above can be found in [10]. From the discussion above, the specimen size should be at least 20 times l_z, or 1660 μm. This calculated value for the specimen size is alarming because McCauley [16] reports typical grain sizes of AlON between 150–250 μm, which based on the arguments presented thus far, suggest fracture in AlON is strength-controlled. Since this cannot be the case, the realistic values for the critical fracture stress, σ_c is probably much higher then the bulk measured flexure strength reported in [16]. For example if $\sigma_c = 800$ MPa, then $l_z = 12$ μm and the specimen size in the finite element calculation

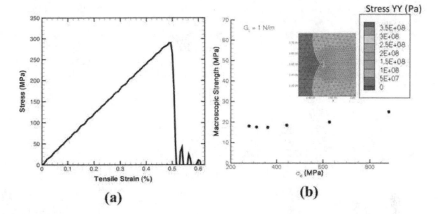

Figure 4: (a) Result of the two grain tension test. The stress increases until approximately 307 MPa, the reported flexure strength of AlON, then the crack propagates along the contact interface. (b) Plot showing simulate fracture process using cohesive interface approach is toughness-controlled. The critical fracture stress, σ_c specified at the grain boundary is increasing but the macroscopic strength remains almost unchanged.

should be 245 μm which falls in the range of typical AlON grain sizes. This analysis provides even greater motivation for conducing micro-testing of AlON bi-crystals. However, note that in real materials there probably exists a distribution of fracture energies at the grain boundaries as a result of grain orientation, defects, etc. There is little data in the literature concerning micro-testing in compression or tension of AlON, and for brittle ceramics in general. Yasuda et al. [27] examined grain boundaries in Al_2O_3 and found observed a distribution of critical fracture energy release rates. There is work ongoing at the Army Research Laboratory to conduct micro experimentation on brittle ceramics.

For the 50 grain simulations, if $\sqrt[3]{50}$ grains per edge are assumed for the computational mesh and the grain size of 542 μm is chosen, then the length of a computational domain is $\sqrt[3]{50} \times 542$ μm = 2.0 mm. While the typical grain size of ALON ranges from $100 - 250$ μm, this large grain size was initially chosen so that the total sample size is on the order of laboratory sized specimens. More realistic grain sizes are chosen for the larger microstructures (note that Paliwal et al. [20] used ALON specimens with dimensions of $5.2 \times 4.2 \times 2.3$ mm for high-rate laboratory testing using a Kolsky bar setup). The elastic longitudinal wave speed of AlON is approximately 10200 m/s, so it takes $(0.002 \text{ m})/(10200 \text{ m/s}) = 196$ ns for a longitudinal wave to cross the specimen, therefore dynamic equilibrium is expected to be reached in approximately 1 μs (which is assumed to be complete when a longitudinal wave crosses the specimen about four times). Specimens were loaded with a range of constant strain rates. During the initial stages of building the three-dimensional computational framework, the microstructures were loaded so that the top boundary of the mesh was moved at a constant velocity of 0.2 m/s (so that the nominal strain rate, $\dot{\epsilon}_{nom}$ = velocity/initial specimen height = $(0.2 \text{ m/s})/(0.002 \text{ m}) = 100$ s^{-1}). Since a sudden velocity impulse is imposed on the microstructure, this corresponds to a shock loading. Various other boundary conditions could be applied as needed, such as confinement loading.

Figure 5 shows the engineering stress-strain response for 50 grain loading in tension and compression. As expected, when loaded in tension the failure strength is approximately 307 MPa. In contrast, the peak compressive strength is 871 MPa at $\dot{\epsilon}_{nom} = 100$ s^{-1}. This is an interesting result because Paliwal et al. [20] measured 720 MPa at qausi-static loading rates ($\dot{\epsilon} = 1 \times 10^{-3}$ s^{-1}).

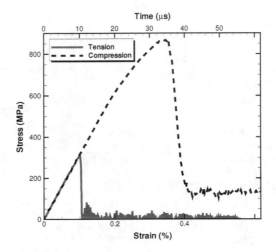

Figure 5: Result of the 50 grain tension test.

Figure 6a shows contour plots of stress in the Y-direction (which is the axis of loading) of a tension simulation at various times. At 6.11 μs the AlON specimen appears to be in a homogeneous stress state. At 7.71 μs, stress concentrations at grain boundaries and triple junctions begin to appear and become more pronounced until 10.1 μs when it seem a crack begins to open, which is very evident 300 ns later (at 10.4 μs) with the blue area of the image representing a location of the microstructure where a crack has propagated and unloaded the material. This is also where the peak strength in tension is observed, as shown in Figure 5. Figure 6b shows an image of the microstructure at 10.4 μs rendered transparent to show how most of the grain boundaries are still carrying load (have high stress, colored red). This is also evident in Figure 6c which shows a cross section of the microstructure also at 10.4 μs.

Figure 7a shows contour plots of stress in the Y-direction (which is the axis of loading) of the compression simulation at various times. At 27.66 μs, stress concentrations at grain boundaries and triple junctions are pronounced until 32.45 μs when a multiple cracks are activated and begin to open. Once cracks are activated and begin to propagate, it is not until 35 μs, or 2.55 μs later that the compressive peak strength is reached. It is interesting to note that at peak stress (which would occur somewhere between 34.58 and 35.64 μs in the shown contour plots) the maximum crack density (all grain boundaries fractured) is not reached. Figure 7b shows an image of the

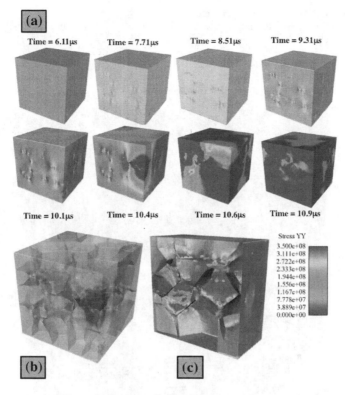

Figure 6: Contours for the 50 grain tension test.

microstructure at 35.64 μs rendered transparent to show how most of the grain boundaries are still carrying load (have high stress, colored red). This is also evident in Figure 7c which shows a cross section of the microstructure also at 35.64 μs.

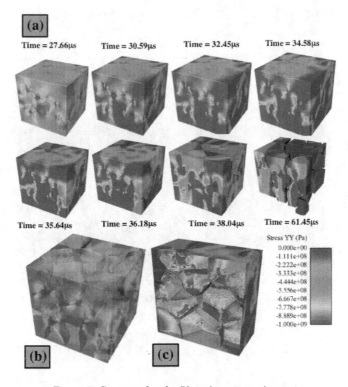

Figure 7: Contours for the 50 grain compression test.

A major thrust of the current research program is to develop the capability to run mesoscale simulations with virtual microstructures with thousands of grains. This is a challenging objective since contact is used to handle grain-grain interactions and does not scale linearly on massively parallel supercomputers. For the 1000 grain simulations, if $\sqrt[3]{1000}$ grains per edge are assumed for the computational mesh and the grain size of 200 μm is chosen, then the length of a computational domain is $\sqrt[3]{1000} \times 200$ $\mu m = 2.0$ mm. The microstructure was loaded so that the top boundary of the mesh was moved at a constant velocity of 0.2 m/s (so that the nominal strain rate, $\dot{\epsilon}_{nom} =$ velocity/inital specimen height = (0.2 m/s)/(0.002 m) = 100 s^{-1}). Unfortunately, the 1000 grain simulation only ran for a few nanoseconds. The reason for this is poor element quality. During the surface-to-volume meshing process very small tetrahedral volume elements were created as a result of small grain facets formed during Voronoi tessellation.

Since the Voronoi tessellation approach for modeling large microstructures did not work, microstructures developed using the Monte Carlo-based grain growth approach (Figure 2c) were used. For the 1220 grain simulations, if $\sqrt[3]{1220}$ grains per edge are assumed for the computational mesh and the grain size of 200 μm is chosen, then the length of a computational domain is $\sqrt[3]{1220} \times 200$ $\mu m = 2.13$ mm. The microstructure was loaded so that the top boundary of the mesh was moved at a constant velocity of 0.2 m/s (so that the nominal strain rate, $\dot{\epsilon}_{nom}$ = velocity/inital specimen height = (0.213 m/s)/(0.00213 m) = 100 s^{-1}). This approach also suffered from some mesh issues, however the simulation stepped far enough in time for some of the grain boundaries to begin to crack under compressive loading as shown in Figure 8a.

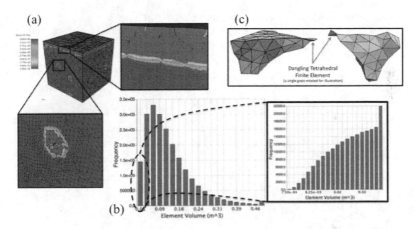

Figure 8: First attempts at modeling 1220 grains in compression.

While the surface meshing procedure produced most surface triangles with acceptable aspect ratios, the surfaces of the grains still had many problematic areas, including non-manifold topology, such as dangling triangles and edges as shown in Figure 8c. In contrast to the 1000 grain mesh that was constructed using Voronoi tessellation, the 1220 grain mesh constructed using the Monte Carlo-based grain growth approach did not have a sudden increase in small volume elements near the left side of the element volume distribution as shown in Figure 8b. A large number of elements with small volumes indicates a poor mesh because the critical time step is controlled by the smallest element length. Instead, simulation problems were caused by contact interactions between dangling tetrahedral elements and other grain boundary junctions.

CONCLUSIONS

Multiscale modeling of ceramics provides valuable tools to understand the physics and mechanics of the brittle material inelastic response. This work was able to demonstrate three-dimensional mesoscale simulations with fracture explicitly represented. Atomistic simulations of AlON were used to calculate the anisotropic response from first principles. Simulations were conducted on AlON, and the size of mesoscale simulations were on the order of laboratory-sized tests. Mesoscopic simulations were successfully able to capture the failure response at the microstructural level and compared well to experimental results reported previously for AlON. Since microstructural topology is included, compression strengthening appears naturally in the mesoscopic failure results. While the work shows great promise and applicability, there are some critical pieces of the framework that need to be refined. As illustrated by simulations conducted on 1000 grains or more the finite element mesh design and quality is extremely important and can cause inability to conduct simulations at the mesoscale. In addition, interfacial grain boundary properties are not known, and were assumed to all be equal to the bulk flexure strength reported in the literature. However, in real materials there exists a distribution of fracture energies at the grain boundaries as a result of grain orientation, defects, etc., therefore, in order to get more accurate results micro-testing of AlON bi-crystals would be extremely helpful. In addition more work needs to be done to develop computational microstructures that are suitable for explicit finite element methods which use the cohesive interface approach. While combining the individual pieces of the multiscale effort into a coherent framework are in the preliminary stages, this work demonstrates the feasibility of the approach and highlights areas where more attention is needed.

References

[1] G. I. Barenblatt. The mathematical theory of equilibrium of cracks in brittle fracture. In H.L. Dryden and T. von Karman, editors, *Advances in Applied Mechanics*, pages 55–129. Academic Press, New York, 1962.

[2] A. Bowyer. Computing dirichlet tessellations. *The Computer Journal*, 24(2):162–166, 1981.

[3] G. T. Camacho and M. Ortiz. Computational modelling of impact damage in brittle materials. *International Journal of Solids and Structures*, 33(20–22):2899–2938, 1996.

[4] J. D. Clayton. Continuum multiscale modeling of finite deformation plasticity and anisotropic damage in polycrystals. *Theoretical and Applied Fracture Mechanics*, 45:163–185, 2006.

[5] D. S. Dugdale. Yielding of steel sheets containing slits. *Journal of the Mechanics and Physics of Solids*, 8:100–104, 1960.

[6] H. D. Espinosa and P. D. Zavattieri. Grain level analysis of crack initiation and propagation in brittle materials. *Acta Materialia*, 49:4291–4311, 2001.

[7] H. D. Espinosa, P. D. Zavattieri, and S. K. Dwivedi. A finite deformation continuum/discrete model for the description of fragmentation and damage in brittle materials. *Journal of the Mechanics and Physics of Solids*, 46(10):1909–1942, 1998.

[8] M. L. Falk, A. Needleman, and J. R. Rice. A critical evaluation of cohesive zone models of dynamic fracture. In *Journal de Physique IV (5th European Mechanics of Materials Conference in Delft, Netherlands)*, pages Pr.5.43 – Pr.5.50, 2001.

[9] E. K. Graham, W. C. Munly, J. W. McCauley, and N. D. Corbin. *Journal of American Ceramic Society*, 71:807–812, 1988.

[10] R. H. Kraft. *Computational Modeling of Damage in Brittle Materials*. PhD thesis, Johns Hopkins University, 2008.

[11] R. H. Kraft and J. F. Molinari. A statistical investigation of the effects of grain boundary properties on transgranular fracture. *Acta Materialia*, 56(17):4739–4749, 2008.

[12] R. H. Kraft, J. F. Molinari, K. T. Ramesh, and D. H. Warner. Computational micromechanics of dynamic compressive loading of a brittle polycrystalline material using a distribution of grain boundary properties. *Journal of the Mechanics and Physics of Solids*, 56:2618–2641, 2008.

[13] A. Kuprat. Modeling microstructure evolution using gradient-weighted moving finite elements. *SIAM Journal of Scientific Computing*, 22(2):535–560, 2000.

[14] G. Lamberton and T. Harnett. Personal communication, 2009. Gary Lamberton –University of Mississippi, Tom Harnett –Surmet.

[15] S. Maiti, K. Rangaswamy, and P. H. Geubelle. Mesoscale analysis of dynamic fragmentation of ceramics under tension. *Acta Materialia*, 53:823–834, 2004.

[16] J. W. McCauley. Structure and properties of AlN and AlON ceramics. In *Elsevier's Encyclopedia of Materials: Science and Technology*, pages 127–132. Elsevier Science, Ltd., 2001.

[17] M. Nygårds. *Microstructural Finite Element Modeling of Metals*. PhD thesis, Royal Institute of Technology, 2003. Doctoral Thesis no. 53.

[18] M. Ortiz. Microcrack coalescence and macroscopic crack growth initiation in brittle solids. *International Journal of Solids and Structures*, 25:231–250, 1988.

[19] M. Ortiz and S. Suresh. Statistical properties of residual stresses and intergranular fracture in ceramic materials. *Journal of Applied Mechanics*, 60:77–84, 1993.

[20] B. Paliwal, K. T. Ramesh, and J. W. McCauley. Direct observation of the dynamic compressive failure of a transparent polycrystalline ceramic (AlON). *Journal of the American Ceramics Society*, 7:2128–2133, 2006.

[21] A. C. Palmer and J. R. Rice. The growth of slip surfaces in the progressive failure of overconsolidated clay. *Philosophical Transactions of the Royal Society of London, Series A*, 332: 527–548, 1973.

[22] M. Radovic and E. Lara-Curzio. Personal communication, 2009. M. Radovic and E. Lara-Curzio– ORNL.

[23] A. D. Rollett and P. Manohar. Monte carlo modeling of grain growth and recyrstallization. In D. Raabe and F. Roters, editors, *Continuum Scale Simulation of Engineering Materials*, chapter 4, page 855. Wiley-VCH, 2004.

[24] D. H. Warner and J. F. Molinari. Micromechanical finite element modeling of compressive fracture in confined alumina ceramic. *Acta Materialia*, 54:5135–5145, 2006.

[25] Z. Wu and J. M. Sullivan. Multiple material marching cubes algorithm. *International Journal for Numerical Methods in Engineering*, 58(2):189–207, 2003

[26] X. P. Xu and A. Needleman. Numerical simulations of fast crack growth in brittle solids. *Journal of the Mechanics and Physics of Solids*, 42:1397–1434, 1994.

[27] K. Yasuda, J. Tatami, T. Harada, and Y. Matsuo. Twist angle dependence of interfacial fracture toughness of (0001) twist boundary of alumina. *Key Engineering Materials*, 2:573–576, 1999.

FUTURE TRANSPARENT MATERIALS EVALUATED THROUGH PARAMETRIC ANALYSIS

Costas G. Fountzoulas and James M. Sands

U.S. Army Research Laboratory, Weapons & Materials Research, APG, RDRL-WMM-B and RDRL-WMM-D MD 21005-5069, USA

Currently used transparent materials applied to military systems are selected by accepting some performance limitations. New polymeric and ceramic materials of improved mechanical and physical properties are continually sought to improve protection and extend service performance for current and future U.S. military transparency systems. The rapid advancement of the computer power and the recent advances in the numerical techniques and materials models have resulted in accurate simulation of the ballistic impact into multi-layer transparent armor configurations. Parametric analysis of existing transparent materials of known materials models can contribute to the development of future materials. The systematic analysis of the effect of material properties, such as modulus of elasticity, yield strength and ultimate tensile strength, on the ballistic performance of armor systems can provide researchers and the manufacturers crucial design information for next generation armor technologies. The ballistic behavior of "future" materials is studied analytically, and by ANSYS/AUTODYN commercial software to develop design goals for properties desirable in next generation materials. The computational results are compared to available experimental data of known materials in order to understand the opportunities from existing markets to improve performances. The goal of the current modeling effort is to develop a correlation between ballistic failure mechanisms and the material properties leading to future materials technologies for soldier protection.

INTRODUCTION

Transparent armor consists either of a single material or a system of materials which are designed to be optically transparent, and to be able to protect from fragmentation or ballistic impacts. The backing has a refractive index closely matching that of the hard face such as to allow substantial transparency of the transparent armor system. The hard face serves to disburse energy caused by the impact of an incoming projectile with the transparent armor system, while the backing serves to retain any pieces of the hard face fractured during ballistic impact. Transparent armor systems using ceramics as the striking face have been explored since the early 1970's because they potentially provide superior ballistic protection to conventional glass based transparent armor systems [1]. Commercially available transparent armor systems are utilized in a variety of military and civilian applications including face shields, goggles, vehicle vision blocks, windshields and windows, blast shields, and aircraft canopies. In addition, transparent armor windows must be compatible with night vision equipment. High performance transparent armor systems typically consist of several different materials, such as polymethyl methacrylate (PMMA), float or soda lime glass and polycarbonate (PC) bonded together with a rubbery interlayer such as polyurethane (PU) or polyvinylbutyral (PVB). Other advanced transparent systems can contain more exotic transparent armor materials such as sapphire, ALON or spinel. The lamination sequence, material thicknesses and bonding between layers have been shown to drastically affect system performance and it has been observed that each material serves an important function. Over the years there have been numerous developmental efforts to optimize the ballistic performance of lightweight, multi-layered transparent armor systems. The majority of these efforts are experimental, which can be time consuming and costly. However, the existing materials technology is 30 years old there is an urgent need for development of new transparent materials combined in proper sequence architectures to face successfully the current more lethal threats. In addition to becoming lighter and having improved mechanical and optical properties, such that the new transparent armor can meet the newer, more stringent multihit specifications. Possible areas of improvement are the manufacturing of a more effective hard face and/or stronger adhesives between the laminate parts and defeat more lethal threats. The U.S. Army has invested heavily in the development of next generation materials, including ceramics, for military systems [2]. The result of the on-going investments is a critical understanding of ceramics strengths and weaknesses for military platforms. This understanding can be applied to the development of next generation transparent armor systems.

Finite element modeling has progressed substantially in its ability to predict failure of materials under extreme dynamic loading conditions. One of the limitations of predictive models is lack of a complete dynamic materials properties database, which is needed for materials models for each of the materials in the simulations. In order to compensate for parameters whose dynamic values were extrapolated from their static or quasi-static properties, baseline experiments are often used to calibrate the models. (3, 4) However, the recalibration method of modeling lacks many of the physical properties and failure mechanisms associated with real-world materials. Therefore, often recalibrated models lack the ability to predict within statistical error future failures over any substantial ranges due to the existence of defects, and materials substitutions often lead to new calibration requirements. The desired approach is to validate a fully characterized materials database with one calibration model, and subsequently apply the model to modified designs. However, despite its apparent problems, recalibration of existing materials models has been proven to be an effective tool in the hands of the modeler by minimizing the number of simulation iterations and resulting in more successful predictions. Regardless of methodology, finite element tools can be applied effectively to reduce the variability between impact tests; and can be used to parameterize designs with fewer experimental failures when robust models are created [3,4].

The objective of this research is to create a notional transparent material by modifying various mechanical properties of a known transparent material, such as modulus of elasticity, and to study its dynamic behavior by comparing it with the dynamic response of the known transparent material.

MODELING

One of the advantages of modeling methods is the ability to create physically challenging architectures to investigate effects of point defects on failure. The sensitivity of ballistic measurement tools is typically less than ±10% due to the range of available failure modes invoked in the high energy exchange between projectile and target. Additionally, capturing the real-time failure modes in the impact event requires highly specialized video equipments. These factors contribute to a very difficult and expensive set of experiments for investigating small flaws and the impact on performance in the experimental realm.

The ballistic behavior of all the targets (Fig. 1), which consisted of float glass, known baseline or modified baseline material and polycarbonate, bonded together by polyurethane, and impacted by a surrogate projectile, was simulated using the non-linear ANSYS/AUTODYN commercial package [5]. The material models used were obtained from the AUTODYN library. The model laminate consisted of panels of spinel and polycarbonate of 232 cm^2 cross sectional area. The thicknesses of the float glass, polyurethane and polycarbonate were kept constant while the thickness of the modified acrylic was varied form 100 to 50 percent of its initial thickness. The projectile applied in the models was a 30 mm long, 1095 steel projectile, of conical frustum geometry, (6-mm large base, 1-mm small base), using two-dimensional axisymmetric models. The solver used for all materials was smooth particle hydrodynamics (SPH). The particle size was set to 0.25 mm. The polycarbonate was simply supported at the corners by applying zero velocity along the x-direction as a boundary condition. The impact velocity was held constant at 878 m/s for all simulations.

The material models used for the float glass, polycarbonate, polyurethane and steel projectile were obtained from the AUTODYN material library [5]. The float glass was modeled using a linear equation of state (EOS), elastic strength model and principal stress failure criterion. The PC was modeled using a shock equation of state (EOS), piecewise Johnson-Cook (JC) strength model, and a plastic strain failure criterion. The projectile steel was modeled using a shock EOS and a JC strength model (5). The baseline material (BLM) was modeled using a shock equation of state (EOS), Von–Mises strength model and a principal stress failure criterion.

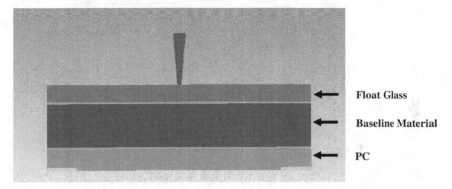

Float Glass

Baseline Material

PC

Figure 1. Typical target architecture

Table 1 shows basic properties of the transparent materials. The exit velocity of all simulations was measured and it was used in determining the properties of a promising future material.

Table 1. Properties of transparent materials

	Modulus of Elasticity (GPa)	Yield Strength (MPa)	Poisson's Ratio ν
Glass	62	-	0.22
Baseline Material Variables	2.74	134.5	0.37
PC	2.76	80.6	0.38

RESULTS/DISCUSSION

The simulation of each model lasted until the projectile had exited the target or its exit velocity

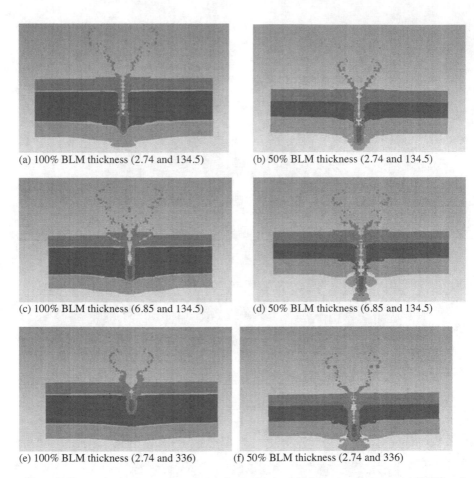

(a) 100% BLM thickness (2.74 and 134.5)   (b) 50% BLM thickness (2.74 and 134.5)

(c) 100% BLM thickness (6.85 and 134.5)   (d) 50% BLM thickness (6.85 and 134.5)

(e) 100% BLM thickness (2.74 and 336)   (f) 50% BLM thickness (2.74 and 336)

Figure 2. Penetration progress of targets consisted of various thickness baseline material (BLM) and varying modulus of elasticity and yield strength after 200 μs simulation time. The first number represents the modulus of elasticity in GPa and the second the yield strength in MPa

was zero. To study the effect of the modulus of elasticity and yield strength on the exit velocity the thickness of the baseline material was decreased by 50 and 75 percent of its original thickness which corresponds to 11 and 22 percent areal density reduction respectively. For each thickness the modulus of elasticity ranged as 2.74, 4.11, 5.48 and 6.85 GPa, and the yield strength ranged as 134.5, 202, 269 and 336 MPa. For each thickness, four separate models were created by combining each modulus of elasticity with all yield strengths. All simulations showed, and it also becomes apparent on Figure 2, that higher stiffness and yield strength increase the ballistic performance of the target; however, the impact of yield strength is more significant. A comparison of Figures 2d and 2f shows higher yield stress retards the projectile exit more than higher modulus of elasticity. Figure 3 shows the damage

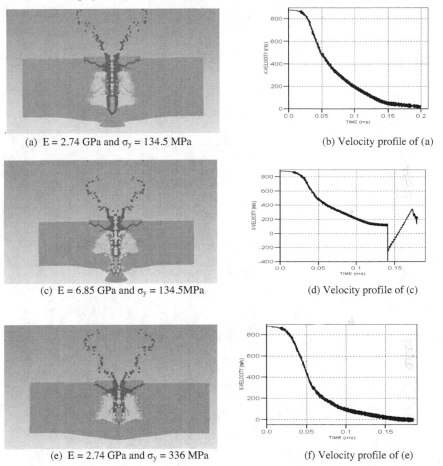

(a) E = 2.74 GPa and σ_y = 134.5 MPa

(b) Velocity profile of (a)

(c) E = 6.85 GPa and σ_y = 134.5MPa

(d) Velocity profile of (c)

(e) E = 2.74 GPa and σ_y = 336 MPa

(f) Velocity profile of (e)

Figure 3. Damage status of targets consisted of baseline material (BLM) of 75% thickness and varying modulus of elasticity and yield strength after 200 μs simulation time. In velocity profile figures the x-axis is "Time" in ms and y-axis is "Exit Velocity" in m/s.

caused by the projectile penetrating a similar architecture target consisting of 75% BLM thickness. Figure 3c and 3e show again that a material with higher yield stress results in higher ballistic efficiency. Figure 3c shows that an increase of the material stiffness has resulted in an usual tensile failure, characteristic of ceramics. While higher stiffness did not stop the projectile from exiting the target, higher yield strength stopped it. By comparing Figures 3a and 3b with Figures 3e and 3f we observe that higher yield strength resulted in preventing the exit of the projectile faster than the low yield strength case.

A summary of the above mentioned results is shown on Figure 4 and Figure 5 and on Tables 2 and 3 for 11% and 22% areal density cases respectively. Table 2 shows the equations of the exit velocity for modulus of elasticity 2.74 and 5.48 GPa as a function of the yield stress for 11 percent areal density reduction. The calculated yield stress necessary to stop a projectile is 269.0 MPa for both moduli of elasticity.

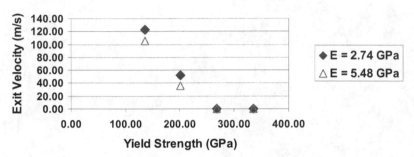

Figure 4. Exit velocity vs. yield strength for 11% areal density decrease

Table 2. Yield strength predictions for projectile stoppage for 11% areal density decrease

Modulus of Elasticity (GPa)	Trend line Equation	Modulus of Elasticity (for zero exit velocity)
2.74	$V = -0.9194\sigma y + 244.12$	269.0 MPa
5.48	$V = -0.7881\sigma y + 206.44$	269.0 MPa

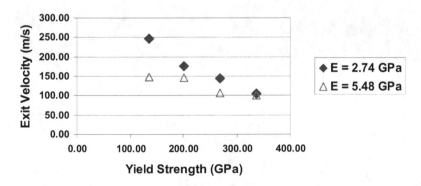

Figure 5. Exit velocity vs. yield strength for 22% areal density decrease

Table 3. Yield strength predictions for projectile stoppage for 22% areal density decrease

Modulus of Elasticity (GPa)	Trendline Equation	Modulus of Elasticity (for zero exit velocity)
2.74	V = -0.684σy + 329.09	481 MPa
5.48	V = -0.276σy + 190.54	690 MPa

Table 3 shows the equations of the exit velocity for modulus of elasticity 2.74 and 5.48 GPa as a function of the yield stress for 22 percent areal density reduction. The calculated yield stress necessary to stop a projectile is 481 and 690 MPa respectively. In this case, it was assumed that the tread line is linear.

CONCLUSION

While the need for advanced materials solutions for protection of vehicles from ballistic threats continues to grow, the ability to predict material performance using advanced modeling tools increases. The current efforts underway in the U.S. Army include the use of ballistic modeling, ballistic testing, and historic knowledge of ballistic design to create structural armors for the transparent armor needs of the U.S. military. The current paper has demonstrated the utility of computational modeling to suggest future desired transparent materials modulus of elasticity and yield strength. The desired modulus of elasticity and yield strength of a new transparent material were determined by modeling of a laminate target consisting of float glass, baseline material, and polycarbonate, which was impacted by a projectile at 878 m/s, for decreased areal density decrease. The analysis showed that

- Areal density decrease is more sensitive to yield strength increase than modulus of elasticity increase
- Projectile defeat with a target having a 11% areal density decrease was shown to be possible for yield strength two times the yield strength of the baseline material
- Projectile defeat with a target having a 22% areal density required yield strength about four times the base line material yield strength to stop the projectile
- Increases in the modulus of elasticity of the baseline material required a corresponding increase in the necessary yield strength for projectile defeat

This is only an initial study, it is noted that further efforts should be placed on the effect of the target architecture, a "new" hard face of improved mechanical properties, and adhesives of stronger chemical bonding to the other laminate parts for further reduction of the areal density.

REFERENCES
[1] Gatti,A & Noone, M J, Feasibility Study for Producing Transparent Spinel (MgAl2O4), AMMRC-CR-70-8, February 1970.
[2] 2006 Army Modernization Plan, "Building, Equipping, and Supporting the Modular Force," Annex D. March 2006.
[3] C.G.Fountzoulas, B.A. Cheeseman, P.G.Dehmer and J.M.Sands, "A Computational Study of Laminate Transparent Armor Impacted by FSP", Proceedings of 23rd Inter, Ballistic Symp., Tarragona, Spain, 14-19 April 2007
[4] C. G. Fountzoulas, J.C.LaSalvia, B.A.Cheeseman, "Simulation of Ballistic Impact of a Tungsten Carbide Sphere on a Confined Silicon Carbide Target", Proceedings of 23rd Inter, Ballistic Symp., Tarragona, Spain, 14-19 April 2007
[5] ANSYS/AUTODYN Vol 11.0, Manual, Century Dynamics Inc., Concord, CA

NANO-PROCESSING FOR LARGER FINE-GRAINED WINDOWS OF TRANSPARENT SPINEL

Andreas Krell, Thomas Hutzler, Jens Klimke, Annegret Potthoff
Fraunhofer Institute of Ceramic Technologies and Systems (IKTS)
01277 Dresden, Germany

ABSTRACT

A clear transparency of components free of scattering losses needs < 0.01% of porosity which is frequently achieved by high sintering temperatures associated with extended grain growth whereas an extreme mechanical stability is bound to fine-grained microstructures < 2 μm. Thus, improvements are expected by technologies which achieve an extreme sintering densification at low temperature. Nano-powders with short diffusion distances do not generally match this target because their high surface curvature gives rise to increased diffusion but is also associated with strong forces of physical interaction (= increasing agglomeration at high specific surface). Therefore, a successful use of such powders for transparent windows needs a high degree of de-agglomeration and processing approaches which provide an extreme homogeneity of the particle coordination in the green shaped bodies. For this end, commercial spinel raw materials from America, Europe and Japan were modified by the manufacturing companies upon specifications of our laboratory in order to promote a "defect-free processing" by an improved dispersibility of the powders. Highly transparent windows up to ~ 240 mm and with thickness > 10 mm were obtained with powders covering a wide surface range of 10-80 m^2/g. However, only successfully de-agglomerated powders with *higher* surfaces ≥ 30 m^2/g were able to achieve a perfect in-line transmission (= density > 99.99%) *at lowest sintering temperatures* resulting in smallest grain sizes (e.g. 270 nm) and a hardness of the spinel on the level of sapphire.

The measured spectral transmission agrees with Mie calculations and an experimentally verified residual porosity in the ppm range.

1. INTRODUCTION: TARGETS OF TRANSPARENT ARMOR DEVELOPMENT
- MICROSTRUCTURAL UND TECHNOLOGICAL CONSEQUENCES

Depending on (i) the ballistic threat and (ii) the backing, the hardness of ceramic armor may or may not contribute to a high ballistic strength.[1] It is, however, a general assumption that in most cases ceramic armor will profit from a high Young's modulus and a high hardness.[2,3,4] With a constant composition and at an identical residual porosity (~ 0 for high transparency) the one tool for increasing the hardness is a minimum grain size of the sintered microstructure. Unfortunately, the impact of the grain size on the hardness of transparent (= zero porosity) spinel ceramics is much smaller than known, for example, for alumina: Fig. 1 shows little difference between the hardness of spinel microstructures with 20 or with 200 μm grain sizes (all with HV10 ~ 12-13 GPa), and only grain sizes in the far sub-μm range provide a spinel hardness which competes with the hardness of sapphire (~ 15 GPa).

The manufacture of optically homogeneous spinel windows with such grain sizes ≤ 0.5 μm needs low sintering temperatures which is a bigger challenge than for opaque sub-μm ceramics because here the achievement of > 99.99% density is an additional extremely important (and difficult) target. Cubic polycrystalline ceramics do not encounter problems with scattering losses associated with bire-fringence as in sintered translucent alumina (Fig. 2). However, here as in all other really transparent materials the residual porosity has to be reduced to less than 0.01% when a high transmittance and a glass-like (not whitish) appearance shall be obtained, and even with such a small porosity significant scattering losses appear at a critical ratio of small (nano-scale) pore sizes scaling with the wave length (Fig. 3a). As a consequence, without any absorption (i.e. at highest purity) scattering losses caused by as few as 0.01% of 40 nm small pores will remove more in-line transmission from the blue than from the red part of the visible spectrum (Fig. 3b) and can, therefore, give rise to some coloration.

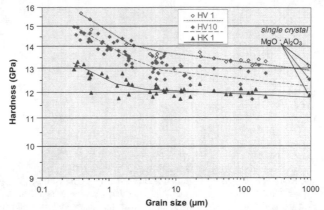

Figure 1 Hardness of sintered transparent (= zero porosity) spinel ceramics with different grain sizes. Knoop (HK1) and Vickers measurements (HV1, HV10) with 1 and 10 kg testing load, respectively; every data point is an average of ≥ 10 individual indents. The data refer to different manufacturing approaches including different commercial spinel powders and spinel powders provided by the authors' inorganic and metallo-organic syntheses[5,6] as well as windows made by reactive sintering.[7] Samples were shaped by cold-isostatic pressing or gelcasting, pre-sintered in air, vacuum, or hydrogen with subsequent hot-isostatic pressing (HIP); comparative samples were densified by hot-pressing. The complete investigations covered the whole range of stoichiometry (MgO·nAl$_2$O$_3$ with n = 1-3).[8]

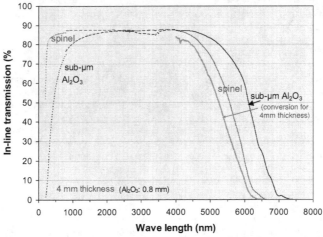

Figure 2. Transmission spectra of sintered transparent grades of spinel compared with corundum (α-Al$_2$O$_3$). The spinel powder was shaped by cold-isostatic pressing, the Al$_2$O$_3$ by slip-casting; both ceramics were manufactured by pressureless pre-sintering in air with subsequent HIP in Ar. Spinel windows with the transmission given here can be obtained both by sintering of spinel powders[5] or by reactive sintering of Al$_2$O$_3$/MgO mixtures.[7] The effect of thickness is described by equ. 1 in reference [9].

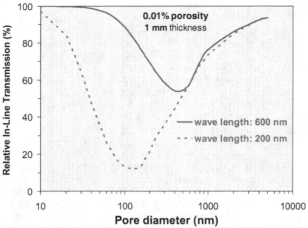

Figure 3a. Calculated effect of the pore-size (Mie scattering) on the in-line transmission of 1 mm thin spinel at 200 and 600 nm wave length (thicker windows are subject to more scattering: Fig. 3b).

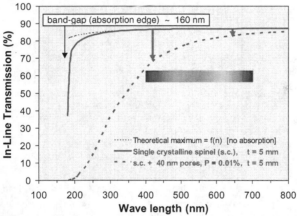

Figure 3b. Calculated Mie scattering losses superimposed to an experimentally measured in-line transmission spectrum of single crystalline spinel with 5 mm thickness.

Another type of "coloration" is grey darkening. High-temperature HIP equipments use graphite heaters (HIP = hot isostatic pressing), and the resulting CO partial pressure gives rise to a reducing atmosphere with the formation of oxygen vacancies in the oxide lattice. The result is a *continuous* depression of the transmission spectrum of transparent spinel starting at far ultraviolet wavelengths (200-250 nm) and covering the whole visible spectrum.[9] The problem is further complicated since it depends not only on the atmosphere but is also influenced by lattice properties as the Al/Mg ratio of spinel.[10,11] This difficulty and measures for minimizing it have been addressed previously[9] thus that the present investigation can compare only such samples where the grey absorption was on a minimum.

In order to investigate the pure influence of different (and differently modified) powders the present study was limited to solid-state sintering without any sintering additives. With this frame and the extreme target of an elimination of last 0.01% of pores at a sufficiently *low* sintering temperature (which keeps the grain size of dense sintered microstructures ≤ 500 nm) very fine-grained powders with a sufficiently high sintering activity are preferred. This aspect seems to favor nanopowders < 100 nm but such powders are also subject to high forces of physical interaction, i.e. agglomeration: going from larger to finer particle sizes the sintering activity first increases towards smaller sizes but deteriorates finally when agglomerates are insufficiently dissolved.[12] Beyond this turning point nano-powders cannot be brought to success by optimizing the sintering procedures. Instead, the key issue is now an optimized dispersion and associated homogeneity of the particle coordination in the green bodies. This development can be focused to optimized powder processing *if*, as for transparent/translucent Al_2O_3 ceramics, commercial powders *are available* which meet such stringent requirements.[5] For transparent spinel ceramics commercial powders are, on the one hand, available with higher specific surfaces > 20 m^2/g whereas, on the other hand, manufacturers' data indicate an increasingly difficult performance the finer these powders are. Therefore, it was the main subject of the present study to investigate how *powder* modifications can contribute to an improved processing of such nanoscale spinel powders when the criterion of success is not a measurable increase of lower densities but the realization of larger windows with a *maximum transmittance* (i.e. density > 99.99%) at smallest grain size.

2. POWDER AND GREEN BODY CHARACTERIZATION
2.1 Characterization of spinel powders (as received and modified)
Tab. 1 summarizes the investigated types of spinel powders and modifications. All powders exhibit a similarly high purity with some small but significant advantage of the coarsest spinel (synthesized from ammonium carbonate). This coarser powder and coarser grades of the other syntheses were investigated with the hope to overcome the agglomeration problems of finer powders < 100 nm; details and micrographs are given in a separate publication.[13] Obvious problems are:

- In the alum-derived spinel powder electron microscopy shows *no individual* particles > 100 nm, but the manufacturer provided a size distribution with a larger median value (d_{50}) of 210 nm. This is a clear indication of agglomeration even in the highly diluted measuring suspension and makes afraid of more severe agglomeration on real processing of more concentrated slurries.
  This fear is the more substantiated by the observation of a substantial solubility of even the coarsest of these high-purity spinel powders in neutral water.[13]
- On the contrary, the flame sprayed spinel (30 m^2/g) contains a significant amount of spherical particles ~ 300 nm, but manufacturer's dynamic laser scattering claimed there were *no* particles > 165 nm.

Table 1. Investigated commercial spinel powders and realized modifications

Powder synthesis	Thermal decomposition of ammonium carbonates	Alum process (static thermal decomp.)	Flame spray pyrolysis
Purity	~ 99.995 % spinel [6-8 ppm Fe]	~ 99.995 % spinel [~ 15 ppm Fe]	~ 99.995 % spinel [10-20 ppm Fe]
Specific surface (BET) → equivalent particle size	13 - 15 m^2/g ~ 120 nm	28 - 31 m^2/g ~ 57 nm	30 - 35 m^2/g ~ 53 nm
Costs (5-20 kg batches) relative to 150nm-α-Al_2O_3 of same purity	~ 3	~ 1.5	~ 30
Modifications	finer by extended milling	de-agglomeration milling coarsened grades:	de-agglomeration milling calcination temperatures coarsened grades:
		14 - 18 - 21 m^2/g	14 - 20 - 25.5 m^2/g

Apparently, there are severe shortcomings both in the *deagglomeration* of the spinel materials (where dissolution in water can promote ionic interactions) and in the known *measuring* approaches:[13]
- Only the coarsest powder (from ammonium carbonate: 13-15 m^2/g) was deagglomerated and measured by common laser scattering without difficulties (Mie evaluation of scattered light confirmed the particle size derived from BET in Tab. 1; equipment at IKTS Dresden: Mastersizer 2000, Malvern).[13]
- However, this Mie evaluation (Mastersizer) comes to its lower bound at about 20 nm particle size which may render measurements of powders < 100 nm increasingly incorrect.
- Therefore, the nano community prefers dynamic laser scattering (DLS = evaluation of Brown's movement; IKTS equipment: Zetasizer Nano ZS, Malvern) which above 1 nm works well for strictly mono-sized powders where the following two difficulties do *not* apply:
  · Using a fluid with low viscosity, the strong dependence of the signal I on the particle size D $(I \sim D^6)$ is not correctly evaluated by the calculating software, and in powders with a wider size distribution the signal of the larger individuals diminishes the contribution of smaller particles.
  · On the contrary, in an organic fluid (minimizing spinel solubility problems) with higher viscosity the larger particles become immobilized and the obtained plot ends at smaller upper sizes than the real distributions.
  As a consequence, most different distributions can be obtained for one powder just depending on the viscosity of the selected fluid. Thus, for 30 m^2/g spinel powders with a common width of the distributions no single viscosity enables the record of the *complete* particle size distribution by DLS whereas both finer and coarser distributions can be produced artificially by tuning the viscosity.[13]

These difficulties were not completely overcome by the present study but some significant improvements have been achieved:

For the as-received flame sprayed spinel, the Mastersizer plot in Fig. 4a is *measured*, without any doubt, correctly with $d_{50} = 2.55\,\mu m$ and almost all "particles" > 1 µm - which means a real and heavy agglomeration. However, with Fig. 4a most of these agglomerates can be dissolved by additional milling, and the one remaining question was whether $d_{50} = 119$ nm is the "real" size of last undestroyed agglomerates (~ 2 * d_{50} of individual particles) or whether this distribution of the milled state was falsified by an increasingly incorrect Mie evaluation of the amounts < 100 nm by the Mastersizer.

Fig. 4b answers this question by a DLS approach where six different "wrong" *partial* distributions were measured at different viscosities. Every of these partial measurements did also deliver an apparent solid loading (of those particles which contributed to this *partial* result), and these apparent solid loadings were used for a final weighted superposition of all these partial results. In Fig. 4b the resulting compound distribution of the flame sprayed powder (derived from the partial DLS data) really shows the expected higher amount of particles < 100 nm than the Mastersizer plot - but, in fact, these two results are very similar and confirm each other. Thus, two hours of attrition milling did *not completely* dissolve *all* agglomerates of the flame sprayed spinel powder but achieved a state where the median agglomerate size was, finally, not larger than about twice the individual particle size.

Only with this confirmation of a reliable measuring approach it became possible to compare powders with a similarly high specific surface but produced by different syntheses: after equal additional milling Fig. 4b shows similar distributions for the flame sprayed and the alum-derived powders with a, however, significantly bigger amount of coarser agglomerates in the latter one.

2.2 Green body manufacture and characterization

A successful milling approach was derived from the best results of de-agglomeration studies (Fig. 4b). Most samples for sintering experiments were prepared by de-agglomeration milling with an additive of about 4% of an organic binder, freeze drying, uniaxial pre-pressing at about 50 MPa and subsequent cold-isostatic pressing at 350 MPa. Regarding recent experimental evidence of a close, governed by technology, correlation of the homogeneity of green ceramic bodies with reduced sinter-

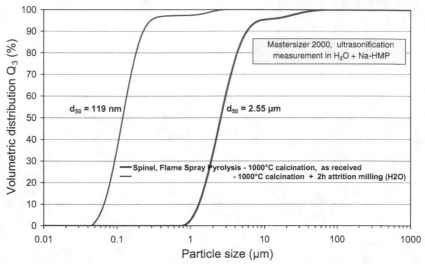

Figure 4a. Particle size distribution of flame sprayed spinel (~ 30 m²/g) after calcination at 1000°C (Mastersizer 2000). As-received state and after 2 h attrition milling.

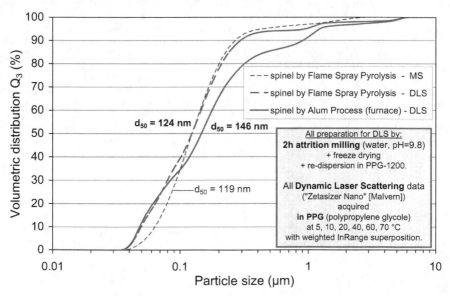

Figure 4b. Comparison of different measuring approaches (DLS and Mastersizer [MS]) and of particle size distributions of flame sprayed and of alum-derived spinel (both ~ 30 m²/g, 2 h attrition milling).
ing temperatures (enabling the highest density at a minimum of grain growth)[14] a few comparative experiments were performed by gelcasting which is known to ensure an excellent homogeneity for transparent alumina.[12] These gelcasting tests were limited to the 14 m²/g spinel.

With cold-isostatic pressing, the different de-agglomeration performance of the two powders with about 30 m²/g from different syntheses (Fig. 4b) was directly transferred to the homogeneity of the green bodies (Fig. 5): the pore size distributions (measurement by mercury porosimetry) show a similar relationship as the particle (more exactly: the *agglomerate*) size distributions in Fig. 4b. Of course, with the 14 m²/g powder the pore size increases proportionally to the particle size of the powder.

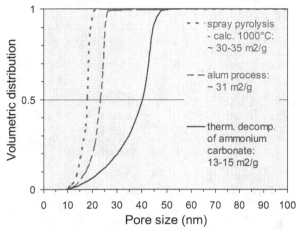

Figure 5. Pore size distributions of green bodies after annealing at 800°C. All samples were shaped by cold isostatic pressing of milled (2 h) spinel powders. See Tab. 1 for basic powder characteristics. Details of the measurement (mercury intrusion) have been published previously.[14]

The porosity was measured both *on the global* level (green density by geometrical and weight measurements) and *microscopically* (Hg intrusion delivers a partial porosity for the evaluated pore size range). The equivalence of both results (→ same ranking in the two columns of Tab. 2) gives evidence of a homogeneous character of these green bodies through the scales. However, the response of these compaction parameters to specific surfaces, particle sizes or processing changes is ambiguous:

Table 2. Porosity and density of CIP-ed green bodies.

Specific surface and origin of powder for green bodies	Microscopic pore volume (by Hg-porosimetry: 9 - 500 nm pores[14])	Total porosity (green density)	
13-15 m²/g; thermal decomposition of ammonium carbonate + 2 h milling		40.3%	(59.7%)
+ 4 h milling	0.159 ml/g	39.3%	(60.7%)
~ 31 m²/g; alum process	0.200 ml/g	44.9%	(55.1%)
30-35 m²/g; spray pyrolysis;			
1000°C calcination	0.172 ml/g	42.9%	(57.1%)
700°C calcination		41.3%	(58.7%)

- As to be expected, the green density was *reduced* (by ~ 4%) if the specific surface increased from about 15 to 30 m²/g (Tab. 2). However,
- processing which improved the dispersibility without such a big increase of the specific surface *increased* the green density, e.g. in Tab. 2 by extended milling of the coarser spinel, by different syntheses (alum process → spray pyrolysis), or by lower calcination (two lots from spray pyrolysis).

This ambiguity diminishes the significance of compaction parameters for a scientific understanding: the correlation of the data in Tab. 2 with the degree of powder dispersion (Tab. 1, Fig. 4b) is poor; it is much better for the pore size distributions of Fig. 5 which reveal additional information about the *local*, microscopic homogeneity of particle coordination.

A *quantitative* evaluation of the merit of the achieved pore size distributions in Fig. 5 comes from the consideration that in a perfect hexagonal-close assembly of monosized spheres the average pore size is about 1/5 of the particle size. Thus, with median particle sizes of 50-60 nm of these 30 m²/g powders an optimum state of homogeneity is represented by a median pore size of 10-12 nm. Regarding the by far *not "mono*sized" character of these commercial spinel powders, the experimentally observed median pore sizes of 17-23 nm are indicative of a "good" but not perfect homogeneity of the particle coordination. The sintering results of section 3 below will confirm this imperfect state by a comparison with the advanced processing of a corundum powder for transparent alumina.

Efforts to improve the degree of homogeneity of green bodies by the use of the coarser, possibly less agglomerated spinel powder derived from ammonium carbonate were not successful: in Fig. 5 bodies made of this 14 m²/g spinel exhibit a median pore size of 40 nm which gives, referred to the particle size, exactly the same "good but not perfect" ratio as observed for the 30 m²/g powders.

Different from the previous experience with alumina,[14] with this 14 m²/g spinel gelcasting did not contribute to a significant improvement of the pore size distribution.

3. SINTERING AND TRANSMISSION RESULTS

Sintering up to densities > 99.99% was performed with undoped samples in three steps:
(i)   Outgassing at 800°C
(ii) followed by pre-sintering up to closed porosity at different temperatures (depending on the powder and its state) for 2 hours (most experiments in air, comparative tests in a vacuum and in different hydrogen atmospheres did not contribute to general improvements). The state of closed porosity was obtained at about 97% relative density but this value fluctuated between 95 and 99% depending on the powders.
(iii) Final densification by HIP between 1250 and 1850°C in Ar with a pressure of 200 MPa.

It is impossible to give all detailed sintering results of the investigated powder modifications here. Instead, Fig. 6 shows some examples and demonstrates how the final results (Tab. 3) were obtained. For every state of a powder it was necessary first to determine the pre-sintering temperature which eliminates the last open porosity *completely*. With the limited accuracy of common density measurements, pre-sintering or HIP temperatures which finally lead to maximum transparency cannot be derived directly from plots as Fig. 6. Instead, for fixing the temperatures which are "optimum" *for transparency* several samples brought to an *apparently* zero open porosity by slightly different pre-sintering treatments were given to different HIP cycles. Only then the transmission measured after HIP reveals *which* of the different pre-sintering states was optimum. The inserts to Fig. 6 give this final fixing of pre-sintering and HIP temperatures.

With this pragmatic schedule "optimum" HIP temperatures were sometimes above and for other powders below the finally fixed pre-sintering temperature. Generally, however, the outlined approach always resulted in very similar temperatures for optimized pre-sintering and subsequent HIP confirming, thus, the previous similar experience with transparent alumina.[15]

CIP-ed samples made of the 14 m^2/g powder (by ammonium carbonate decomposition) showed a significant but small decrease of the sintering temperature when the milling process was extended from 2 to 4 hours; consequently, their grain sizes were similar after HIP. However, despite the slightly lower pre-sintering temperatures after 4 h of milling the optimum RIT transmission *of these* samples was achieved by a slightly *higher* HIP temperature than that which resulted in the highest transmission after 2 h of milling (Fig. 6 and the following Tab. 3 always give this combination of pre-sintering and HIP temperatures which resulted in the highest RIT transmission).

Remarkably, longer milling of this 14 m^2/g powder enabled a big increase of the transmission - in spite of the small progress in the pre-sintering curves (Fig. 6). On the other hand, neither the sintering temperatures nor the transmission were improved further by gelcasting of this powder.

A stepwise further decrease of sintering temperatures and final grain sizes accompanied by a *steadily improving transmission* was observed with increasing specific surfaces of the spinel powders up to 31 m^2/g (non-optimized alum-derived powder) and, finally, to 81 m^2/g (spinel by IKTS Dresden[6]).

Without the optimized (flame sprayed) spinel with ~ 30 m^2/g this sequence of sintering results in Fig. 6 would suggest a successful elimination of agglomeration in high-surface powders up to 81 m^2/g. And really, in comparison with the non-optimized 31 m^2/g spinel this 81 m^2/g spinel enables what is expected for successful dispersion: a dramatically decreased sintering temperature and a very small final grain size (0.5 μm) enabled, obviously, by the smaller particles of this nanoscale powder.

However, a comparison of this 81 m^2/g grade with the optimized 30 m^2/g spinel demonstrates *still lower* sintering and *less* grain growth *of the coarser* particles *if* these are better de-agglomerated. What is equivalent, on the other hand, with a lower, less perfect dispersibility of the 81 m^2/g spinel.

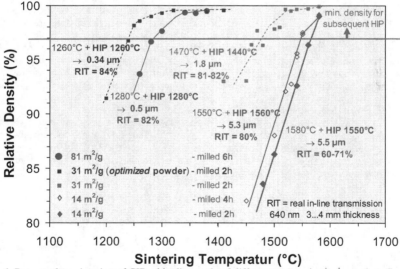

Figure 6. Pressure-less sintering of CIP-ed bodies made of different undoped spinel powders: final optimum (for the transmittance) pre-sintering and HIP conditions, and average grain size and real in-line transmission RIT (measured with a narrow aperture of about 0.5° at 640 nm) of these optimized discs.

Two features are noteworthy in Fig. 6 if the sintering performance of these spinel powders is compared with the advanced state of transparent/translucent Al_2O_3 ceramics[5]:
- The spinel powder with about 14 m²/g (by ammonium carbonate decomposition) needs ≥ 1480°C for full densification to high transparency whereas a similarly processed and CIP-ed corundum powder with a same specific surface and obtained by a similar synthesis achieves the same sintering state at much lower 1300-1330°C after CIP, at 1180-1240°C after gelcasting, and at 1150°C after slipcasting.[5] This big temperature difference is hardly explained by the only small differences of the ionic/covalent nature and of the associated different melting temperatures of spinel (~ 2135°C) and Al_2O_3 (~ 2065°C) and indicates, instead, a lower degree of de-agglomeration and of homogeneity of the green spinel bodies (cp. spinel pore size distributions in Fig. 5 above with corundum results of Fig. 4a in ref.[5]).
- This conclusion is also supported by the observation that CIP-ed bodies manufactured of the "optimized" spinel with 30 m²/g or of the 81 m²/g powder exhibit just *slightly lower* temperatures for sintering to transparency (-35...-50°C) than possible with CIP-ed 14 m²/g Al_2O_3 - a modest progress for an increase of the specific surface by a factor between 2 and 6. Again the comparison indicates a lower degree of homogeneity of even the "best" spinel samples here.

A hypothetical explanation could be the addressed above partial solubility of spinel in water. However, replacing the aqueous by non-aqueous processing approaches did never improve the results significantly in the present study.

Therefore, regarding the challenging example of the excellent performance of a 14 m²/g corundum powder[5] it was tried to improve the homogeneity of the green spinel bodies by asking the commercial manufacturers (alum process and flame spray pyrolysis) for powder grades with somewhat decreased specific surfaces and, by this means, improved dispersibility.

Alum-derived powders with 14, 18 and 21 m²/g all reproduced the good transparency also obtained with the 30 m²/g grade - but only if sintered at *higher* temperatures than the ~ 1470°C which had to be used for the 30 m²/g powder. Sintering of flame sprayed spinel powders with reduced specific surfaces (14, 20, 25.5 m²/g) turned out even more difficult: compared with the 30 m²/g grade of the same synthesis these coarser lots consisted of a mixture of big spheres (up to 1 μm) and a multitude of nanoscale particles with the consequence of *higher* "optimum" sintering temperatures and, nevertheless, *lower transmission* values. In fact, both alum-derived and flame sprayed spinel powders with about 14 m²/g all requested similar temperatures of 1530 ± 50°C for pre-sintering and HIPing to transparency as observed for the ammonium carbonate derived spinel with this surface area. With all these grades it was impossible to realize the lower (by 200-300°C!) sintering temperatures known for the manufacture of transparent/translucent Al_2O_3 from a 14 m²/g corundum powder.

Tab. 3 shows the results for "best" grades from the different syntheses. At similar specific surface of about 30 m²/g, the relatively modest advantage of the flame sprayed powder in its dispersibility (Fig. 4b) and in the homogeneity of the green bodies (Fig. 5) is accompanied by a tremendous decrease in the sintering temperatures of these *undoped* (!) samples and enables a remarkable improvement of the microstructure and of some of the optical and mechanical (cp. Fig. 1) parameters.

Table 3. Sintering of undoped spinel powders after cold-isostatic pressing. See Fig. 1 for hardness data (HK1, HV1, HV10) of transparent windows with the grain sizes given here.

Powder synthesis	Thermal decomposition of ammonium carbonates 13 - 15 m²/g	Alum Process (static thermal decomp.) 28 - 31 m²/g	Flame spray pyrolysis 30 - 35 m²/g
Optimum temperatures for pre-sintering + HIP	1480...1580°C	1440...1500°C	1260...1280°C
Real in-line transmission at thickness t ~ 4 mm	~ 80%	81-82%	~ 84% (~ 81 % at t = 6 mm)
Typical av. grain size	3 - 6 μm	0.8 - 2 μm	0.3 - 0.6 μm

The transmission curves in Fig. 7 confirm the ranking of the spinel powders given by Tab. 3 for the visible part of the spectrum and provide additional insight:

First, similar transmission results can be obtained by both gelcasting or by CIP and are frequently independent of the HIP temperatures over a wide range.

More interesting in Fig. 7 are the differences at the transition from the visible to the ultraviolet part of the spectrum: the finer the original state of the powder, the stronger the tendency of a decreasing transmission at shorter wave lengths. With the result of exact Mie calculations in Fig. 3b it can be suggested that this increasing susceptibility of windows *made of the finest* powders to *deteriorated* transmission at short wave lengths is caused by *nano*scale pores which survive the HIP process just in *such* windows. This hypothesis is not readily compatible with common sintering physics where it is assumed that these *smallest* pores should be eliminated *first*: their extreme surface curvature and small size give rise to locally highest surface diffusion over short distances during solid state sintering. Experimental evidence is, however, very clear:[13] independently of the sintering and HIP-ing temperatures transmission and high-resolution scanning electron microscopy confirm the existence of

- 45-200 ppm of 45-70 nm small pores at triple junctions of windows made of spray pyrolysis spinel powder ($\sim 30 \, m^2/g$) by HIP between 1260-1650°C, whereas only
- 4 - 18 ppm of these pores are present in transparent discs made of $14 \, m^2/g$ spinel grades (by ammonium carbonate or alum decomposition) HIP-ed between about 1500-1750°C.[13]

A tentative explanation of this unexpected feature could be the much higher *number* of pores in green bodies made of the finest powders: a high number of small pores could increase the statistical probability that some of them survive the HIP process *in spite of* their small size (high surface curvature).

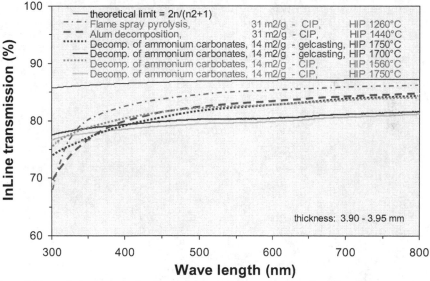

Figure 7. Transmission spectra of undoped spinel windows made of powders from different syntheses.

A similar phenomenon is the different optical homogeneity (measured by interferometry at Carl Zeiss SMT as $\Delta n_{633\,nm}$ = root mean square of local wave front deviations / thickness; Δn was calculated after subtraction of Zernike coefficients 1-36 for long-wave effects). As a reference, optical sapphire (for laser applications, 12-15 mm thick) was reported to exhibit Δn = 0.5-0.9 ppm at a wave length of 632 nm.[16] With a thickness of 6-10 mm the present spinel discs are optically *more homogeneous* (smaller Δn) when fabricated from the *coarser* 14 m²/g spinel (by ammonium carbonate decomposition); the homogeneity of these sintered spinel samples is on a similar level with sapphire:

- alum-derived spinel powder (~ 14 m²/g) / CIP $\qquad\qquad$ Δn = 1.3 - 1.5 ppm,
- flame sprayed spinel powder (~ 30 m²/g) / CIP $\qquad\qquad$ Δn = 0.4 - 2.5 ppm,
- alum-derived spinel powder (~ 30 m²/g) / CIP $\qquad\qquad$ Δn = 0.6 - 1.7 ppm,
- spinel by thermal decomposition of ammonium carbonate (~ 14 m²/g) / CIP $\quad$ Δn = 0.4 - 1.1 ppm,
$\qquad\qquad\qquad\qquad\qquad\qquad\qquad\qquad\qquad\qquad$ / gelcasting $\quad$ Δn = 0.24- 0.36 ppm.

This advantage of gelcast spinel samples is a surprise since the defect-avoiding profit of gelcasting (demonstrated previously for a 14 m²/g corundum powder[12,14]) was not confirmed in the present characterization of green spinel bodies and in their pre-sintering performance. It can be speculated that the homogeneity advantage provided by gelcasting *of these spinel* powders is too small for clear records of Hg porosimetry or of density data obtained by Archimedes' measurements. It is, however, not clear how such small homogeneity differences can become significant for the final HIP densification.

The more complete elimination of last pores during HIP of green bodies made of the coarser powders (~ 14 m²/g) and the eventually improved optical homogeneity of such samples initiate the question whether further progress can be achieved with *still coarser* raw powders (then, probably, at higher sintering temperatures). Fig. 8 confirms this idea for spinel experiments with special powders synthesized and processed by IKTS Dresden (commercial spinel powders are not available with surfaces < 13 m²/g *and* a sufficiently high sintering activity). The close matching of *total* and *in-line* transmissions of discs made of 9 m²/g raw powders indicates that the improvement is achieved by minimizing *scattering* losses (and not by eventual changes in the absorption), and the progressing increase of the transmittance towards *shorter* wave lengths confirms that it is a population of *small nano*pores which is eliminated *more* efficiently when starting with a surface of 9 instead of 14 m²/g. The benefit, however, extends then over the whole visible spectrum.

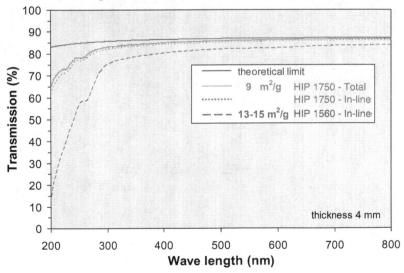

Figure 8. Transmission improvement of undoped spinel windows by use of coarser raw powders.

4. DISCUSSION AND CONCLUSIONS

Generally, the extreme density > 99.99% of transparent ceramics can be achieved by *high*-temperature sintering (accompanied by grain growth to 30-300 μm)[17,18,19,20] or at *low* temperature limiting the grain growth to ≤ 0.5 μm with benefits e.g. for the mechanical performance[5]. Until present it was, however, not clear which mechanisms enabled the surprising decrease of sintering temperatures down to < 1300°C in some of the recent reports.

With Fig. 6 and Tab. 3 it becomes clear that fine powder particles < 100 nm may contribute to this progress but are inefficient if this small particle size is not associated with two other conditions: an optimum degree of de-agglomeration (Fig. 4b) and an extreme homogeneity of particle coordination in the green bodies (Fig. 5). For smaller samples as described in a previous report,[5] deficits in these technological steps and parameters can be balanced e.g. by a high CIP pressure up to about 1000 MPa what is, however, impossible when larger windows have to be shaped in larger CIP vessels limited to a pressure of less than 400 MPa. Additionally on sintering, defects in the green bodies become more critically when the absolute value of shrinkage increases with the component size. Thus, the requirements for optimized powder processing and for an extreme homogeneity of the green bodies become the more stringent the larger the components and the lower the applicable pressures are. The approach described by the present investigation matches these conditions in a way that low-temperature sintering without any doping additives (Fig. 6) resulted in a minimum average grain size of 270 nm combined with a high transmission (Tab. 3, Fig. 9a). The largest tiles approximated a letter-size format (Fig. 9b).

Without doping additives, the most fine-grained windows (Fig. 9a) were HIP-ed starting from modified commercial spinel powders with a specific surface of about 30 m²/g (average particle size ~ 50-60 nm). The optical quality of these windows was good but was further improved when *coarser* raw powders with about 14 m²/g were used. Surprisingly, the best transmission spectrum was achieved by raw powders with a specific surface as low as 9 m²/g (Fig. 8).

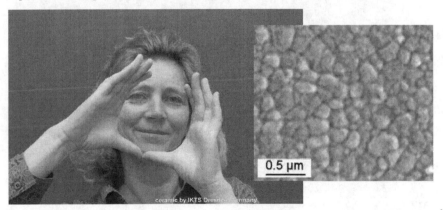

Figure 9a. Transparency and microstructure of a window HIP-ed at 1260°C of an optimized 30 m²/g spinel powder without doping additives. Typical thickness as requested for transparent armor (Fig. 9 in

ref.[5]: 2-3 mm). Average grain size 270 nm (determined as 1.56 * average intercept length[21]). See Tab. 3 for grain sizes and transmission data obtained with other spinel powders.

Unfortunately, the coarser spinel powders did (without doping additives) never enable similarly low pre-sintering and HIP temperatures $\leq 1300°C$ as it is common for manufacturing advanced translucent/transparent Al_2O_3 from a 14 m^2/g corundum powder.[5] With spinel, the specific surface had to be increased by more than 100% (≥ 30 m^2/g) for similarly low sintering temperatures (Fig. 6). It is, therefore, speculated that the dispersibility of the investigated spinel powders (Fig. 4b) and the present - fairly good - homogeneity of the shaped green bodies (Fig. 5) are not perfect yet and should be subject to further improvement.

A consequence of the high "optimum" (for maximum transmission) HIP temperatures of e.g. ~ 1530°C for 14 m^2/g spinel powders is that fine-grained microstructures < 2 μm were not obtained with these coarser powders. Therefore, within the frame of the present study windows with high transparency *and* grain sizes ≤ 0.5 μm enabling a hardness on the level of sapphire (HV10 = 14.5 - 15 GPa) were manufactured starting from powders with specific surfaces > 25 m^2/g only.

The grain size-hardness correlation for transparent spinel windows manufactured from different raw materials by most different processing and sintering regimes is given by Fig. 1 for several measuring approaches and testing loads. This complex body of data (including the full range of spinel homogeneity $MgO \cdot nAl_2O_3$ with n = 1-3 [8] and the comparison of undoped HIP with LiF-doped hot-pressing) does not confirm the existence of transparent spinel grades with a microhardness of 16-17 GPa (e.g. measured as HV0.5 at 20-30 μm grain size[20]) as claimed by a few reports.

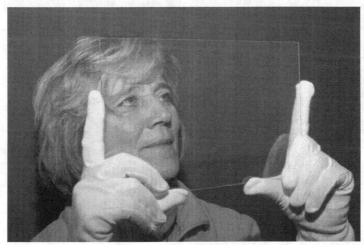

Figure 9b. Transparent window with 240 mm length, thickness 2 mm. Alum-derived powder HIP-ed without doping additives at 1480°C.

The present investigation conflicts with the simplifying idea that nanoparticles would generally sinter at lower temperatures: this expectation can be matched only as far as the increasing with the specific surface agglomeration is dissolved successfully, and physical reasons set a lower bound for this target. This critical view to yet finer particles is supported by other studies with HIP as with hot-pressing:

- In comparison with the present results achieved by the use of undoped spinel powders of ~ 30 m^2/g (~ 55 nm), *smaller* flame-sprayed spinel particles (~ 33 nm) were reported to require a *much higher* HIP temperature of 1700°C, and the maximum transmittance was relatively low (77% for 2 mm thin samples).[22]
- By hot-pressing, much better results were obtained with a 24 m^2/g spinel derived from metallo-organic precursors than with the 77 m^2/g equivalent of same synthesis but lower calcination.[19]

Therefore, on the presently most advanced level of spinel de-agglomeration a particle size of ~ 55 nm (~ 30 m^2/g) appears as the lower limit for which a maximum transmission can be associated with HIP temperatures < 1300°C and with resulting grain sizes < 500 nm. In fact, the results reported by Fig. 6 for the 81 m^2/g spinel from IKTS Dresden are the one known example of a powder with this high surface sintered below 1300°C to an RIT transmittance of 82% (at 3 mm thickness) with 0.5 µm grain size - but even this result does not compete with results displayed by the Figs. 6, 7 and 9 for a coarser optimized spinel powder with only ~ 30 m^2/g.

ACKNOWLEDGEMENT

Parts of this investigation were performed within the frame of projects with Carl ZEISS SMT, Oberkochen, Germany and with Bundeswehr WTD91, Meppen, Germany.

REFERENCES

[1] A. Krell. E Strassburger, Ballistic strength of opaque and transparent armor, Ceramic Bull. 86 [4, Exclusive Online Articles] 9201-9207 (2007).

[2] J.J. Swab, Recommendations for determining the hardness of armor ceramics, *Int. J. Appl. Ceram. Technol* 1 [3] 219-215 (2004).

[3] N.V. Davis, X.-L. Gao, J.Q. Zheng, Design, characterization and evaluation of material systems for ballistic resistant body armor: a comparative study, pp. 1-23 in: Proc. 49th AIAA/ASME/ASCE/AHS/ ASC Structures, Structural Dynamics, and Materials Conference, The American Institute of Aeronautics and Astronautics, Reston (VA), 2008.

[4] G. Subash, S. Maiti, P. Geubelle, D. Ghosh, Recent advances in dynamic indentation fracture, impact damage and fragmentation of ceramics, *J. Am. Ceram. Soc.* 91 [9] 2777-2791 (2008).

[5] A. Krell, T. Hutzler, J. Klimke, Advanced spinel and sub-µm Al2O3 for transparent armor applications, *J. Europ. Ceram. Soc.* 29 [2] 275-281 (2009).

[6] K. Waetzig, A. Krell, J, Klimke, Synthesis of re-dispersible high-purity nanoscale spinel, spinel powder and sintered transparent spinel products, unpublished German Patent Application (Oct. 27, 2009).

[7] K. Waetzig, T. Hutzler, A. Krell, Transparent spinel by reactive sintering of different alumina modifications with MgO, *cfi/Ber. Dt. Keram. Ges.* 86 [6] E47-E49 (2009).

[8] A. Krell, A. Bales, Grain size dependent hardness of transparent magnesium aluminate spinel, submitted to *Int. J. Appl. Ceram. Technol.*

[9] A. Krell, J. Klimke, T. Hutzler, Transparent compact ceramics: Inherent physical issues, *Optical Mater.* 31 [8] 1144-1150 (2009).

[10] J. Kim, Effect of ZrO2 addition and nonstoichiometry on sintering and physical property of magnesium aluminate spinel, PhD thesis, Case Western Reserve University, Cleveland (OH), 1992.

[11] C.-J. Ting, H.-Y. Lu, Defect reactions and. the controlling mechanism in the sintering of magnesium aluminate spinel, *J. Am. Ceram. Soc.* 82 [4] 841-848 (1999).

[12] A. Krell, T. Hutzler, J. Klimke, Transmission physics and consequences for materials selection, manufacturing, and applications, *J. Europ. Ceram. Soc.* 29 [2] 207-221 (2009).

[13] A. Krell, T. Hutzler, J. Klimke, A. Potthoff, Larger fine-grained components of transparent spinel by different nanopowders, submitted to *J. Am. Ceram. Soc.*

[14] A. Krell, J. Klimke, Effect of the homogeneity of particle coordination on solid state sintering of transparent alumina", *J. Am. Ceram. Soc.* **89** [6] 1985-1992 (2006).

[15] A. Krell, P. Blank, H. Ma, T. Hutzler, M.P.B. van Bruggen, R. Apetz, Transparent sintered corundum with high hardness and strength, J. Am. Ceram. Soc. **86** [1] 12-18 (2003).

[16] J. Dong, P. Deng, Ti:sapphire crystal used in ultrafast lasers and amplifiers, *J. Crystal Growth* **261** [4] 514-519 (2004).

[17] G. Gilde, P. Patel, P. Patterson, D. Blodgett, D. Duncan, D. Hahn, Evaluation of hot pressing and hot isostatic pressing parameters on the optical properties of spinel, *J. Am. Ceram. Soc.* **88** [10] 2747-2751 (2005).

[18] I.E. Reimanis, H.-J. Kleebe, R.L. Cook, A. DiGiovanni, Transparent spinel fabricated from novel powders: Synthesis, microstructure and optical properties, SPIE Defense and Security Symposium Proceedings (Orlando/FL 13.-15.4.2004), The Society of Photo-Optical Instrumentation Engineers, Bellingham (WA), 2004. Electronic document available at:
http://www.tda.com/Library/docs/Spinel_paper%20DoD%20EM%20Symp_022.pdf.

[19] R. Cook, M. Kochis, I. Reimanis, H.-J. Kleebe, A new powder production route for transparent spinel windows: Powder synthesis and window properties, pp. 41-47 in: SPIE Defence and Security Symposium Proceedings, Vol. 5786, *IX. Window and Dome Technologies and Materials* (Orlando/FL, 28.3.2005). Edited by R.W. Tustison, The Society of Photo-Optical Instrumentation Engineers, Bellingham (WA), 2005.

[20] J.L. Sepulveda, R.O Loutfy, S. Chang, Defect free spinel ceramics of high strength and high transparency, *Ceram. Eng. Sci. Proc.* **29** [6] 75-85 (2008).

[21] M.I. Mendelson, Average grain size in polycrystalline ceramics", *J. Am. Ceram. Soc.* **52** [8] 443-446 (1969).

[22] A. Goldstein, A. Goldenberg, Y. Yeshurun, M. Hefetz, Transparent $MgAl_2O_4$ spinel from a powder prepared by flame spray pyrolysis, *J. Am. Ceram. Soc.* **91** [12] 4141-4144 (2008).

EXPERIMENTAL METHODS FOR CHARACTERIZATION AND EVALUATION OF TRANSPARENT ARMOR MATERIALS

E. Strassburger and M. Hunzinger
Fraunhofer Institut für Kurzzeitdynamik (EMI)
Am Christianswuhr 2
79400 Kandern, Germany

J.W. McCauley and P. Patel
U.S. Army Research Laboratory
AMSRD-ARL-WM-MD
Aberdeen Proving Ground, MD 21005

ABSTRACT
Visualization techniques are essential for the observation and understanding of the mechanisms of interaction between projectiles and target materials. At the Fraunhofer Institute for High-Speed Dynamics, Ernst-Mach-Institute (EMI), different methods have been developed, which allow the observation of either damage propagation or projectile penetration in transparent materials and laminates. An Edge-on Impact (EOI) test method coupled with a high speed Cranz-Schardin camera, with frame rates up to 10^7 fps, has been developed to visualize damage propagation and dynamic fracture in structural ceramics.

Since the view on the crater zone and the tip of a penetrating projectile is obscured by damaged material, a flash X-ray technique has to be applied in order to visualize projectile penetration. The flash X-ray cinematography method, which allows recording eight radiographs in a single test, will be presented.

The latest development at EMI to observe ceramic fragments is a Laser-Lightsheet illumination technique, coupled with a high-speed video camera, which allows observing fragments ejected from the impact crater during projectile penetration. With this method it is possible to determine velocity and size of the ceramic ejecta as a function of time. Experimental results with all methods and their correlation to ballistic resistance will be discussed.

INTRODUCTION
Transparent armor is one of the most critical components in the protection of light armored vehicles. In many cases the occupants the vehicle, who are visible through the windows, are the main target of direct fire attacks. The windows have to provide protection not only against infantry ammunition, but also against fragments and blast waves from detonations. Typical transparent armor consists of several layers of glass with polymer interlayers and backing. The design of transparent laminates for ballistic protection is still an empirical process mainly. Considering the high number of parameters influencing the performance like number, thickness and type of the glass layers, thickness and type of the bonding layers and the polymer backing, the necessity to have tools for a systematic optimization becomes obvious. The field of parameters is even extended with an additional front layer of transparent ceramic, which has been proved to enhance the efficiency of transparent laminates against armor piercing ammunition significantly.

Adjusting all the components of a laminate in an efficient way requires an understanding of the mechanisms of projectile penetration into the different materials. This process can be divided into two parts, projectile deformation and erosion on one hand and target material damage and failure on the other hand. Since part of transparent armor consists of brittle materials, the fragmentation of the ceramic and glass layers plays a key role in the resistance to penetration. Shockey et al. described the failure phenomenology of ceramic targets impacted by long rod projectiles[1], and recently the failure of glass[2] due to the penetration of steel projectiles of size and shape similar to the steel cores of armor piercing ammunition.

The data derived from the post penetration analysis of the fragmentation were used as a basis for modeling material failure and projectile penetration[3]. In these models the resistance to penetration is mainly attributed to residual strength, determined by friction and flow characteristics of the failed material. However, the analysis of fragmentation after penetration is completed reveals the final state of the projectile-target interaction, but cannot deliver direct information on the dynamics of the process.

The development of novel techniques for visualization of projectile-target interaction has been the focus of several research projects at EMI for many years. Already established for a longer time, but constantly adapted to new applications, is the so-called Edge-On Impact (EOI) Test. The EOI technique has been developed at EMI in order to visualize dynamic fracture in opaque and transparent brittle materials[4]. Edge-On Impact tests allow a characterization of ceramic materials by the macroscopic fracture patterns, single crack velocities and crack front velocities (damage velocities). This technique can be applied during the shock wave phase of the interaction to characterize material failure due to stress waves, prior to the actual penetration.

Since the view on the interaction zone is quickly getting obscured by damaged material, flash X-ray techniques have to be applied in order to visualize projectile penetration. At EMI the flash X-ray cinematography method, which allows recording eight radiographs in a single test, has been developed. With this technique the deformation, erosion and penetration of the projectile can be observed and quantitatively analyzed.

Although the material ahead of the impacting projectile has failed and is pervaded by a very dense network of cracks, all the fragments are still in their original place during the first phase of interaction. The fractured material is also under high pressure, exerted by the projectile. Thus, there are no density changes on a macroscopic scale in the fractured material which means, that material failure cannot be detected with flash X-ray techniques at that time. Since the material which is in direct contact with the projectile or in the immediate vicinity cannot be visualized inside the target, an experimental method was developed which allows observing the fragments ejected from the crater during penetration. The key to the observation of single particles in the dense cloud of ejecta is the Laser-Lightsheet illumination technique, coupled with a high-speed video camera. With this method it is possible to determine velocity and size of the ceramic ejecta as a function of time.

The different visualization methods presented in this paper enable the observation of target and projectile material failure through the different phases of interaction. Examples of the application of the techniques are presented in the following sections.

EDGE-ON IMPACT

In an Edge-On Impact (EOI) Test the projectile hits one edge of the specimen and fracture propagation is observed by means of a high-speed camera. At EMI a Cranz-Schardin camera with frame rates up to 10^7 fps has been developed for this purpose. Two different optical configurations are usually employed. A regular transmitted light shadowgraph set-up is used to observe wave and damage propagation and a modified configuration, where the specimens are placed between crossed polarizers and the photo-elastic effect is utilized to visualize the stress waves. Pairs of impact tests at approximately equivalent velocities are then carried out in transmitted plane (shadowgraphs) and crossed polarized light. Figure 1a) shows a schematic of the Edge-on Impact test with the added crossed polarizers; Figure 1b) illustrates an exploded view of the impactor/sample interaction.

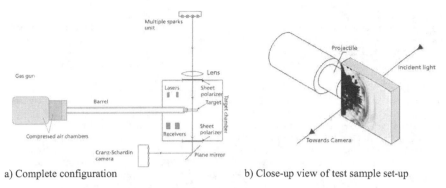

a) Complete configuration   b) Close-up view of test sample set-up

Figure 1. Schematic of EOI Test set-up[5]

Both steel solid cylinder (mass 127 g) and spherical impactors (mass 39.1 g) have been used at impact velocities from 270- 925 m/s to determine baseline data on monolithic plates of dimensions 100x100x10 mm of different glass, glass ceramic and ceramic materials. Most runs were carried out at ~ 400 m/s. The data collected from the EOI test consists of a series of 20 photographs as a function of time, typically at 0.25 - 2 µs intervals. Pairs of impact tests at approximately equivalent velocities are carried out in plane and crossed polarized light to correlate the dynamic fracture with the associated stress fields. Detailed graphs are then created plotting crack, damage and compression and shear stress wave velocities.

Experiments performed in plane light show the evolution of damage and material failure, while the photoelastic visualization illustrates the stress wave propagation as a function of time. Figure 2a shows a selection of two shadowgraphs (top) and corresponding crossed polarizers photographs (bottom) of a baseline test with Starphire® high-purity soda-lime glass (PPG, Pittsburgh, PA), impacted by a spherical steel projectile of 16 mm diameter at 440 m/s.[5, 6]

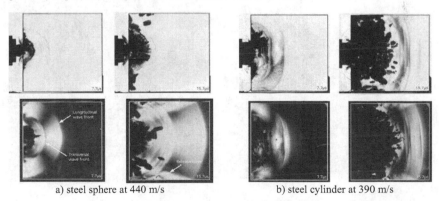

a) steel sphere at 440 m/s   b) steel cylinder at 390 m/s

Figure 2. Selection of two shadowgraphs (top) and crossed pol. photographs (bottom) from impact on Starphire glass[5]

The shadowgraphs show a crack front growing from the impacted edge of the specimen. Only one crack center can be observed close to the upper edge of the specimen. The crossed polarizers photographs illustrate the propagation of the longitudinal and the transversal stress waves. Release waves due to reflections at the upper and lower edge can also be recognized. Note that damage appears dark on the shadowgraphs and the zones with stress birefringence are exhibited as bright zones in the crossed polarizers photographs.

Figure 2b shows a selection of two shadowgraphs along with the corresponding crossed polarizers photographs of the baseline tests with the cylindrical projectile. A coherent damage zone is growing from the impacted edge, preceded by a zone with separated crack centers, probably initiated by the stress waves interacting with defects. It can be recognized that the stress wave front appears more advanced and exhibits a different curvature in the crossed polarizers view. This seeming discrepancy can be explained by the different sensitivities that the different optical techniques employed exhibit with respect to the stress level that can be visualized. In a shadowgraph image the light intensity depends on the second spatial derivative $\partial^2 n/\partial x^2$ of the refractive index, whereas in the crossed polarizers set-up the intensity of the transmitted light depends on the photo-elastic properties of the material. Therefore, it is possible that the first visible wave front in the shadowgraph configuration appears at a different position than the forefront of the stress wave, visible in the crossed polarizers set-up. Both techniques can visualize different parts of the same stress wave.

Crack and damage patterns, very similar to those observed with soda-lime glass, can also be recognized with other types of glass[7] and glass ceramic[8]. Figures 3 and 4 illustrate wave and damage propagation in fused silica[9] and TRANSARM (ALSTOM), a transparent lithium alumino silicate glass ceramic.

Recent progress in material technology has also made available aluminum oxynitride (AlON) as a polycrystalline ceramic that fulfills the requirements of transparency and requisite mechanical properties for transparent armor. Figure 5 compares the fracture patterns in AlON when impacted with a steel sphere (a) and a steel cylinder (b). In contrast to the shadowgraphs, where a wave front is not discernible, the crossed polarizers configuration reveals an approximately semicircular wave front which for the cylindrical impactor is only a little further advanced compared to the damage front visible in the shadowgraphs at the same time.[10]

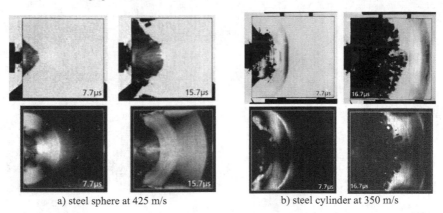

a) steel sphere at 425 m/s   b) steel cylinder at 350 m/s

Figure 3. Selection of two shadowgraphs (top) and crossed pol. photographs (bottom) from impact on Fused Silica

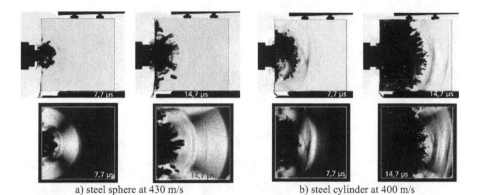

a) steel sphere at 430 m/s   b) steel cylinder at 400 m/s

Figure 4. Selection of two shadowgraphs (top) and crossed pol. photographs (bottom) from impact on TRANSARM glass ceramic

When the spherical projectile hits the edge of the AlON specimen, a fan-shaped fracture pattern can observed, consisting of many single cracks, emanating from the center of impact, which has been theoretically studied by Grinfeld et al. (2007)[11].

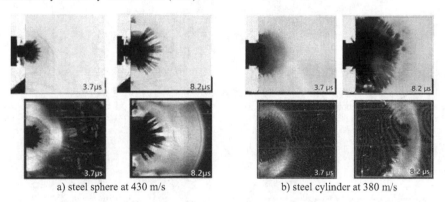

a) steel sphere at 430 m/s   b) steel cylinder at 380 m/s

Figure 5. Selection of two shadowgraphs (top) and crossed pol. photographs (bottom) from impact on AlON polycrystalline transparent ceramic

Different types of cracks are generated and different fracture velocities can be observed at one impact velocity in one specimen. Therefore, it is necessary to distinguish between the velocity of single, continuously growing cracks and crack/damage fronts. The term damage velocity is used here to denote the velocity of the fastest fracture which was observed in a ceramic material. In order to determine the damage velocity the distances of fracture tips and/or the fracture front are measured and plotted versus time. Linear regression of the data delivers fracture and fracture front velocities. Figure 6 shows the path-time history of wave and damage propagation from the baseline test with

Starphire glass[6], presented in Figure 2b. In addition to wave propagation, the expansion of four crack centers at the front of the damage zone was also analyzed. The slope of the straight line through the nucleation sites, which is denoted damage velocity v_D, was 3269 m/s, which means, that the damage velocity was close to the transversal wave velocity.

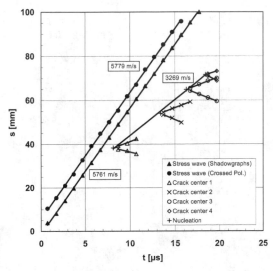

Figure 6. Path-time history of wave and damage propagation in Starphire® glass

Table I shows a compilation of crack and damage velocities of two types of glass, fused silica, a glass ceramic and a polycrystalline transparent ceramic. In each of the ceramics damage velocity is a function of impact velocity. The damage velocities approach the longitudinal wave velocity c_L at high loadings. Therefore, damage velocities are compared for a constant impact velocity of about 400 m/s in Table I.

Table I. Crack and Damage velocities determined from EOI-tests

Material		Crack velocity (m/s)	Damage velocity (m/s)
Starphire	(soda-lime glass)	1580	3270
Borofloat	(borosilicate glass)	2034	4150
Fused Silica		2400	5121
TRANSARM	(glass ceramic)	2151	4950
AlON	(polycrystalline ceramic)	4377	8381

The visualization of the onset and development of damage during the interaction of the projectile with the target material offers a unique opportunity for direct comparison of experimental data to model predictions. The EOI-technique has therefore been used as a tool for the validation of damage models by several researchers[12, 13, 14, 15].

Since transparent armor can consist of many layers, the thickness and type of bonding layers can significantly influence the performance of the system. The influence of a polyurethane (PU) bonding layer on wave and damage propagation was investigated in an EOI-configuration with cylindrical projectiles[5]. Tests with specimens consisting of two parts of the dimensions 50 mm x 100 mm x 9.5 mm were conducted in order to examine the influence of interlayer thickness. Figure 7 illustrates a comparison of wave propagation and damage in Starphire® specimens with bonding layers of thickness 0.64 mm, 2.54 mm and 5.08 mm. The impact velocity was 380 ± 5 m/s in all tests. The upper line of pictures shows the shadowgraphs, while the corresponding crossed polarizers photographs are presented in the lower line of pictures, respectively. Figure 7a illustrates the specimens at 10.7 µs and Figure 7b at 23.7 µs after projectile impact. The shadowgraphs at 10.7 µs show that the first glass layer (left part of specimen) is damaged through the coherent fracture front growing from the impacted edge and through the nucleation of crack centers, initiated by the longitudinal stress waves. At that time, no damage can be recognized in the second glass layer (right part of specimen). The crossed polarizers photographs demonstrate, that the first longitudinal stress pulse has not yet crossed the thickest PU bonding interlayer (right), whereas the stress wave is clearly visible in the right half of the specimens with the thinner glue interlayer.

| 0.64 mm | 2.54 mm | 5.08 mm | 0.64 mm | 2.54 mm | 5.08 mm |

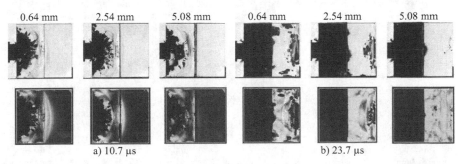

a) 10.7 µs　　　　　　　　　　　　　　　b) 23.7 µs

Figure 7. Starphire laminates with interlayer of different thickness impacted by steel cylinder at 380 m/s[5]

After 23.7 µs (Figure 7b) the compressive stress pulse has already been reflected as a tensile wave at the rear edge of the specimens in all three cases. The shadowgraphs illustrate that damage in the second glass layer is mainly due to the tensile wave and starts from the rear edge of the specimen. In the case of the thickest glue interlayer only little damage was observed in the second glass layer.

The EOI technique has recently been applied to investigate a variety of interface geometrics including saw tooth, corrugated and wave shaped. In addition, the influence of layers of strengthened glass has been studied[16].

FLASH X-RAY CINEMATOGRAPHY

The penetration of a high-speed projectile into a target material can only be visualized by means of flash-radiography. For this purpose, usually several flash X-ray tubes are arranged around the target and the radiographs are recorded on X-ray film. A simple method is the multi-exposure of one film or alternative detector. This method can only be applied when the number of objects is limited and the objects can easily be distinguished. The upper limit of the number of projections is set by the saturation of the detector due to multi-exposures.

A different approach is realized by a set of geometrically separated channels and an array of slits, in order to prevent multi-exposure. Due to geometrical boundary conditions with respect to the target set-up and safe distances the number of channels is also limited. Therefore, both methods allow only pseudo cinematography of the process to be observed, since the radiographs of several experiments have to be combined in order to get a time-resolved image of the process. However, this requires a high reproducibility of the experiments, which can be difficult to achieve in a series of tests. The lower the reproducibility, the higher is the number of tests needed. For this reason it is desirable to have a flash X-ray system that provides a high-number of radiographs in just one experiment. Such a system has been developed at EMI[17].

A schematic of the measurement set-up for flash X-ray cinematography is shown in Figure 8. Instead of several separate X-ray tubes one multi-anode tube is utilized. In the multi-anode tube eight anodes are arranged on a circle of ≈ 12 cm diameter. This configuration causes only a relatively small parallax for the projections from the different anodes. The process under observation can be X-rayed at eight different times. The radiation transmitted through the target is then detected on a fluorescent screen. The position of the target is between the multi-anode tube and the fluorescent screen, relatively close to the fluorescent screen. The fluorescent screen converts the radiograph into an image in the visible wavelength range, which is photographed by means of an intensified digital high-speed camera. The maximum frame rate that can be achieved with such a system depends on the decay time of the fluorescent screen, the time characteristics of the intensifier and the camera. Frame rates of 100 000 fps have been achieved with a fast decaying fluorescent screen and have been used in this study.

The new X-ray cinematography technique was applied in order to study the dwell-penetration transition with small caliber AP projectiles impacting ceramic/glass/polycarbonate targets[18]. The phenomenon of dwell with small caliber AP projectiles at impact velocities below 1000 m/s was already observed during the pioneering studies of Wilkins[19], who examined the interaction of 7.62 mm AP projectiles and surrogate steel penetrators with thin ceramic/aluminum targets. Using the classic flash X-ray technique Wilkins observed, that the steel projectiles did not penetrate the ceramics during the first ≈ 20 μs after impact. During this phase the projectiles were eroded to about half of their initial length. The phenomenon, that a projectile does not (or only very little) penetrate a target over a period of time is designated as dwell. Several studies with small calibre projectiles on the dwell phenomenon[20, 21] have demonstrated that erosion or "wear" of the steel core is one key factor in the energy dissipation of the projectile and thus, for the ballistic resistance.

In order to improve the protective strength of ceramic composite targets it is important to know the relations between the duration of dwell, the type and thickness of the ceramic material and the influence of the backing. Time resolved flash X-ray cinematography is a powerful tool for studying the influence of the different parameters on dwell and the ballistic resistance, since the number of tests can be significantly reduced compared to classic flash X-ray techniques.

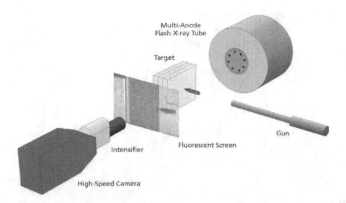

Figure 8. Schematic of typical flash X-ray cinematography set-up[18]

The penetration of a 7.62 mm armor piercing projectile with steel core into a ceramic/ glass/ polycarbonate target consisting of a 4 mm front layer of 0.6 μm grain size, full density sintered Al_2O_3 (Fraunhofer IKTS), three borosilicate glass layers of 9 mm thickness, respectively, and a 3 mm polycarbonate plate is illustrated in Figure 9. The 20 flash X-ray images of Figure 9 originate from three tests, performed with equal target set-ups and impact conditions. The frame rate in each single test was 10^5 fps. The times of the X-ray flashes were shifted from test to test. In the first test the flash times were set to 3 μs, 13 μs,..., 3 μs + n·10 μs, with n = 1, 2,3,...,7. In the second test the flash times were 5 μs,..., 5 μs + n·10 μs, and n·10 μs in the third test. The projectile did not penetrate the first glass layer until 23 μs after impact. During the interaction with the ceramic front layer the length of the projectile was significantly reduced by erosion. The strong effect of the ceramic layer on the projectile is remarkable, particularly with regard to the small size (30 mm x 30 mm) and thickness of the ceramic tile. In order to visualize the interfaces between the single glass layers, thin copper foils were inserted between the glass layers. Due to the higher X-ray absorption in the copper the interfaces appear dark.

The position-time histories of the projectile tail, nose and the target rear surface are shown in Figure 10. The results from the different tests are represented by different symbols in Figure 10, whereas the tail, nose and target rear surface positions are distinguished by different shades of grey. The surface of the ceramic front layer is at the zero position on the ordinate. The path-time histories indicate that only little penetration occurred during the first 10 μs. After this dwell phase a penetration velocity of about 400 m/s was observed. During the time interval of observation the penetration velocity decreased to about 220 m/s. The projectile was stopped with this target configuration.

About 40 μs after impact a strong deformation of the glass layers seems to begin. It is obvious, that a brittle material like glass cannot support such deformations without failure. Since the density changes in the material associated with crack formation are too small and the resolution of the X-ray imaging system is not sufficient, failure cannot be visualized with this technique. However, failure and damage evolution in transparent armour can be visualized by means of high-speed photographic methods.

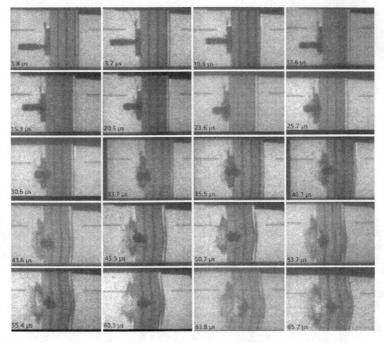

Figure 9. Flash X-ray cinematography of 7.62 AP projectile penetrating a ceramic/glass/polycarbonate laminate; images compiled from three tests[18]

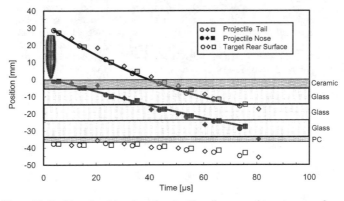

Figure 10. Position-time histories of projectile tail, nose and target rear surface from tests presented in Figure 9

LASER-LIGHTSHEET TECHNIQUE

When a high-speed projectile hits a ceramic the material ahead of the projectile is severely fragmented. Very soon after the onset of penetration ceramic fragments are being ejected from the crater. If it is assumed that the ejected fragments in the immediate vicinity of the projectile have been in contact or close to the projectile tip, an analysis of the ejecta could provide information on the state of fragmentation of the ceramic in front of the projectile. Data of the variation in time of the fragment size could even clarify the dynamics of the process. Figure 11 illustrates the formation of such a fragment cloud with a sequence of six selected high-speed photographs from the penetration of a 7.62 mm steel core projectile into an Al_2O_3-ceramic tile of 4 mm thickness on an aluminum backing.

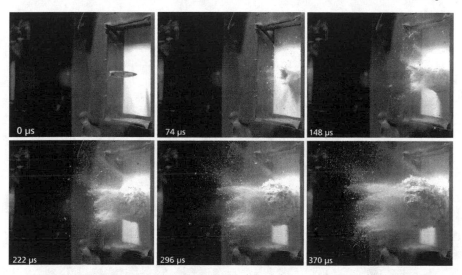

Figure 11. Selection of high-speed photographs from penetration of a 7.62 mm AP projectile into 4 mm Al_2O_3-ceramic on aluminum backing

During the first 150 µs the ejection of fragments nearly perpendicular to the ceramic surface can be observed. Later, the formation of a second cone of fragments, including bigger particles, can be recognized around the inner fragment zone. The high-speed photographs demonstrate the high density of particles, especially in the beginning, which prevents a quantitative analysis of the particles by conventional optical observation techniques.

Thus, in order to establish an instantaneous time resolved correlation between the ejected fragments and the penetration process, a novel optical method was designed. The light-sheet technique with the adaption of a Laser represents a non invasive method to visualize single particles in a defined measuring plane with a high time resolution rate. It allows determining speed, direction of motion and size of single particles. Figure 12 shows a schematic of the experimental set-up.

The punctiform Laser-beam is lead into a special light-sheet-optic and converted into to a linear divergent beam. The light segment of about 1 mm thickness is led by a mirror from the top of the target box in front of the ceramic. The orthogonally to the ceramic's surface orientated light-sheet defines the measurement plane in which the particles are illuminated during the experiment. Particles outside the measurement plane are not or only weakly illuminated. Additionally the depth of focus of the camera which is arranged orthogonally to the light-sheet has to be as small as possible and exactly adjusted to the illuminated plane. This way it is assured that only the light scattered from the fragments placed in the plane of the light-sheet are displayed with a clear cut to the image plane. If there is a high density of particles, the fragments out of the measurement plane are significantly weaker illuminated and only appear as a fog, clearly distinguishable from the fragments staying directly in the light-sheet. To avoid undesirable reflections and therefore additional illumination of particles outside the measurement plane the light sheet has to be coupled out through a glass panel into a beam catcher. Figure 13 shows the typical assembly in the laboratory.

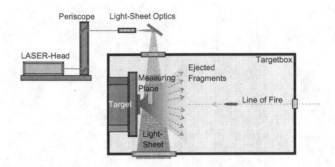

Figure 12. Schematic of the Laser-light-sheet illumination technique

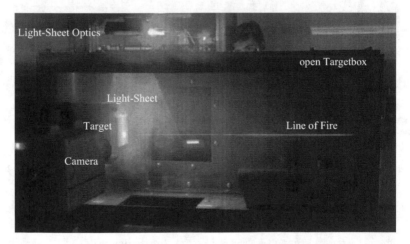

Figure 13.   Photograph of laboratory set-up: open target box with artificial fog

With an artificial fog, both the light-sheet and the line of fire are visible in the open target box. The position of the high-speed camera in the photograph is the same during the experiment, whereas the target box is sealed during the test with a sidewall. Taking into consideration the possible frame rate and the image resolution of the used CMOS-camera, the measured area (rectangular zone within ejected fragments area, Figure 12) must be restricted to a small, but significant array. Assuming a statistically symmetric distribution of the fragments within the cone of ejecta, the measurement plane was accomplished as an elongated rectangle above the line of fire. Thus, it was possible to detect all particles in the measurement plane above the line of fire, including those with high initial velocities of more than 300 m/s. Since the particles were visible in two or more consecutive photographs the velocities could be determined. Figure 14 shows a sequence of 10 photographs from a high-speed video which was recorded at a rate of 100,000 frames per second. The series of photographs from right to left clearly visualizes several particles of different size. The target consisted of 6 mm spinel of 0.6 μm grain size, backed by an aluminium plate and its position was about 4 cm from the edge of the high-speed photographs. The lower edge of the high-speed photographs corresponded to the line of fire. With the settings chosen here, it was possible to discriminate particle sizes from about 170 μm up to some millimetres. Bigger particles which were not completely located in the light-sheet, were only partially illuminated. In these cases only an outline of the fragment was visible. If particles have a sufficiently high component of velocity orthogonal to the light-sheet plane, they can appear or disappear without crossing the edges of the high-speed photographs.

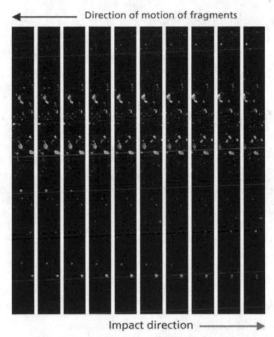

Figure 14. Sequence of 10 high-speed photographs from video recorded with Laser light-sheet technique; original size of observed area: 3.1 mm x 50 mm; impact on 0.6 μm grain size spinel

Because of the high number of particles whose fly by is scanned this way, a special tracking-software is needed to support the overall analysis. The software must be able to recognize and distinguish single particles, to determine their dislocation in consecutive photographs and to process the whole sequence of photographs, which usually consists of several hundred per shot.

Figure 15 shows the average fragment size as a function of time, determined from impacts on sintered Al_2O_3 (10.6 μm grain size, density 98.5 %, hardness 1457 HV10) and sapphire (hardness 1522 HV10) ceramic tiles of 4 mm thickness on an aluminum backing. The data in Figure 15 represent a moving average which means that each data point is the mean value from the analysis of 50 consecutive images. The comparison between the two materials shows that, particularly during the first 500 μs, the average size of the ejected fragments from the penetration of the sintered Al_2O_3 is clearly bigger than in the case of sapphire. The sintered Al_2O_3 also exhibited a significantly higher resistance in a series of ballistic tests[22]. This result indicates the importance of a time-resolved fragment size analysis which allows for recognizing those ceramic fragments that interacted with the projectile and therefore, recording the dynamics of the ceramic/projectile interaction.

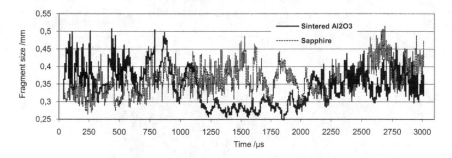

Figure 15. Example for average fragment size versus time plot for sintered Al_2O_3 and sapphire

CONCLUSION

Three complementary visualization techniques have been presented which allow for observing target and projectile material damage and deformation over all phases of interaction. In particular for transparent materials the Edge-on Impact technique enables insights into the onset and development of damage and offers a unique opportunity for direct comparison of experimental data to damage model predictions.

Visualization of the penetration phase can only be accomplished with X-ray techniques. The application of the flash X-ray cinematography method to the case of ceramic faced transparent armor has demonstrated the possibilities to visualize projectile and target deformation and to determine penetration-time histories.

In order to analyze the fragmentation of ceramic target materials during projectile penetration the Laser light-sheet illumination technique has been adapted to visualize the fragments ejected from the interaction zone.

The application of all the methods described here in combination with modeling offers a strong basis for the analysis of projectile-target interaction and therefore, the design of materials with enhanced ballistic resistance and the optimization of targets.

ACKNOWLEDGMENTS
   All the Edge-On Impact work reported here was performed under contracts from the European Research Office supported by the U. S. Army Tank Automotive Research, Development and Engineering Center and the Army Research Laboratory.

REFERENCES

[1] D.A. Shockey, A.H. Marchand, S.R. Skaggs, G.E. Cort, M.W. Burkett and R. Parker, Failure Phenomenology of Confined Ceramic Targets and Impacting Rods, Int. J. Impact Engng., 9, 263-275 (1990)

[2] D.A. Shockey, D. Bergmannshoff, D.R. Curran, and J.W. Simons, Physics of Glass Failure During Rod Penetration, Advances in Ceramic Armor IV, Ceramic Engineering and Science Proceedings Volume 29, Issue 6, 23-32, 2008

[3] D.R. Curran, D.A. Shockey, and J.W. Simons, Mesomechanical Constitutive Relations for Glass and Ceramic Armor, Advances in Ceramic Armor IV, Ceramic Engineering and Science Proceedings Volume 29, Issue 6, 3-13, 2008

[4] E. Strassburger, Visualization of Impact Damage in Ceramics using the Edge-on Impact Technique, Int. J. of Applied Ceramic Technology, Vol. 1, no. 3, 235-242, 2004

[5] E. Strassburger, P. Patel, J.W. McCauley and D.W. Templeton, Wave Propagation and Impact Damage in Transparent Laminates, Proc. of the 23rd Int. Symp. on Ballistics, Tarragona, Spain, 1381-1388, 2007

[6] E. Strassburger, P. Patel, J.W. McCauley, C. Kovalchik, K.T. Ramesh, D.W. Templeton, High-Speed Transmission Shadowgraphic and Dynamic Photoelasticity Study of Stress Wave and Impact Damage Propagation in Transparent Materials and Laminates Using the Edge-on Impact (EOI) Method, 25th Army Science Conference, November 27-30 2006, Orlando, FL, USA

[7] H. Senf, E. Strassburger, H. Rothenhäusler, Stress Wave Induced Damage and Fracture in Impacted Glasses", J. de Physique IV, Colloque C8, Vol. 4; Proc. of EURO DYMAT '94, 741-746, 1994

[8] H. Senf, E. Strassburger, H. Rothenhäusler, A Study of Damage During Impact in ZERODUR; J. Phys. IV, Colloque C3, Vol 7, Proc. of EURODYMAT 97, C3-1015-1020, 1997

[9] E. Strassburger, P. Patel, J.W. McCauley and D.W. Templeton, High-Speed Photographic Study of Wave and Fracture Propagation in Fused Silica, Proc. of the 22nd Int. Symp. on Ballistics, Vancouver, Canada, 761-768, 2005

[10] E. Strassburger, P. Patel, J.W. McCauley and D.W. Templeton, Visualization of Wave Propagation and Impact Damage in a Polycrystalline Transparent Ceramic - AlON, Proc. of the 22nd Int. Symp. on Ballistics, Vancouver, Canada, 769-776, 2005

[11] M.A. Grinfeld, J.W. McCauley, S.E. Schoenfeld, T.W. Wright, Failure Pattern Formation in Brittle Ceramics and Glasses, Proc. of the 23rd Int. Symp. on Ballistics, Tarragona, Spain, 953-963, 2007

[12] S. Hiermaier, W. Riedel, Numerical Simulation of Failure in Brittle Materials using Smooth Particle Hydrodynamics; Proc. of Int. Workshop on New Models and Numerical Codes for Shock Wave Processes in Condensed Media, Oxford, UK, 1997

[13] C. Denoual, F. Hild, Dynamic Fragmentation of Brittle Solids: A Multi-Scale Model, European Journal of Mechanics A/ Solids 21, 105-120, 2002

[14] M. Grujicic, B. Pandurangan, N. Coutris, B.A. Cheeseman, C. Fountzoulas, P.Patel, E. Strassburger, A Ballistic Material Model for Starphire®, a Soda-Lime Transparent-Armor Glass,' Materials Science and Engineering A, **491**, 397-411, 2008

[15] M.O. Steinhauser, K. Grass, E. Strassburger, A. Blumen, Impact Failure of Granular Materials – Non-Equilibrium Multiscale Simulations and High-Speed Experiments, International Journal of Plasticity **25**, 161-182, 2009

[16] E. Strassburger, P. Patel, J.W. McCauley, D.W. Templeton, A. Varshneya, High-Speed Photographic Study of Wave Propagation and Impact Damage in Novel Glass Laminates, Proc. of the 24th Int. Symp. on Ballistics, 548-555, 2008

[17] K. Thoma, P. Helberg, E. Strassburger, Real Time-Resolved Flash X-Ray Cinematographic Investigation of Interface Defeat and Numerical Simulation Validation, Proc. of the 23rd Int. Symp. on Ballistics, Vol. 2, 1065-1072, 2007

[18] E. Strassburger, Ballistic Testing of transparent armour ceramics, Journal of the European Ceramic Society **29** (2009) 267-273

[19] M.L. Wilkins, Third Progress Report on Light Armor Program. UCRL-50460, Lawrence Livermore Laboratory, Livermore, CA, USA, 1968

[20] W.A. Gooch, M.S. Burkins, P. Kingman, G. Hauver, P. Netherwood and R. Benck, Dynamic X-ray Imaging of 7.62-mm APM2 Projectiles Penetrating Boron Carbide. Proc. 18th Int. Symp. on Ballistics, Vol. 2, 901-908, 1999

[21] C.E. Anderson, M.S. Burkins, J.D. Walker and W.A. Gooch, Time-Resolved Penetration of B_4C Tiles by the APM2 Bullet, Computer Modeling in Engineering & Science, Vol.8, No. 2, 91-104, 2005

[22] E. Strassburger, M. Hunzinger, A. Krell, Fragmentation of Ceramics under Ballistic Impact, Proc. of 25th Int. Symposium on Ballistics, Beijing, China, May 17-21 2010

METHOD FOR PRODUCING SIC ARMOR TILES OF HIGHER PERFORMANCE AT LOWER COST

Bhanu Chelluri, Edward Knoth and Edward Schumaker
IAP Research Inc.
Dayton, Ohio, USA

Lisa P. Franks
TARDEC, US Army
Warren, Michigan, USA

ABSTRACT

Under an Army funded project IAP has developed an alternate manufacturing technology of Dynamic Magnetic Compaction (DMC) for ceramic powder materials. This technology, based on principles of pulsed magnetic forces, combines sub millisecond compaction process with pressure-less sintering to reduce cycle times by nearly an order of magnitude for ballistic tile fabrication. The tiles processed by the DMC process exhibit properties such as density, hardness, modulus and grain size comparable to that of hot pressed SiC-N material. We achieved a high pressing throughput rate greater than 1350 lbs/day using the DMC method with an estimated cost below $35 per 4" x 4" x 0.75" tile (< $25/lb) once the system is scaled for production.

INTRODUCTION

This paper describes a unique dynamic net shape processing method of ceramic materials such as silicon carbide. Currently high performance ballistic tiles are produced using hot pressing (Pressure Assisted Densification- PAD) process or Hipping (HIP). Both of these processes are batch processes, have low throughput and are expensive due to inherent time and secondary processes involved in producing the final tile shape. Hot pressed SiC material exhibits full density and fine microstructure, which are beneficial for obtaining better mechanical properties and ballistic performance. Conventional pressing with pressure-less sintering has not been able to match density levels and fine grain size of hot pressed material. The pulse/dynamic pressing approaches can be beneficial for accomplishing the above goals as they yield high green density and thus require shorter sintering cycles and retain fine grain structures. In addition, the dynamic magnetic compaction (DMC) approach offers pressing times of sub millisecond duration which, when combined with short sintering cycles, will shorten the production times.

EXPERIMENTAL WORK

In dynamic magnetic compaction (DMC) processing (1), kinetic energy is imparted to powder material using magnetic fields in a sub millisecond time duration. DMC compaction pressures of a few GPa range on powders produces high green compact density. The DMC technique for ductile metallic materials has been demonstrated (2). However, in the case of brittle powders, the compacted material develops cracks due to internal rebound energy. IAP Research, Inc. has developed an innovative design of a dynamic compaction system to control rebound energy and process high performance brittle ceramics. Prior to designing of the actual system for ceramic tiles, finite element modeling of the compaction dynamics was conducted. The focus of such analysis was on the rebound issues pertaining to compacting brittle powder materials. Based on modeling results, iterations of design were carried out and compaction trials conducted to verify the results. Based on the success of such experiments, the system was designed such that three tiles could be pressed in one compaction.

By combining Dynamic Magnetic Compaction (DMC) and the pressure-less sintering approach, we processed SiC armor tiles. These DMC tiles exhibit uniform and high sinter density (>

98-99%) with a lower shrinkage rate than conventional pressure-less sintered SiC material. The hardness, modulus and micro structural properties of the tiles were similar to those of hot pressed material. Initially tiles of 0.75" x 0.75" x 0.4" thickness were produced on a subscale system and then the system scaled up to produce 1.5" x 1.5" x 0.75" thick tiles for ballistic testing

Powders of 490 NDP (ready to press SiC powders) from Superior Graphite Co were used for the dynamic compaction study. The powders were made using spray dry process with boron and carbon as additives. The powder particle sizes were typically between 0.7 - 1.0 micron. The as-received particles were agglomerated and ranged in particle sizes between 50- 250 microns as determined via Clias dry particle size analysis. The as-received powders were used without additional sieving.

HIGH GREEN DENSITY

The compressibility curves of as-received powders were generated using a conventional mechanical press (at University of Georgia Tech) to understand the powder response. Figure 1 shows the compressibility curve of 490 NDP using conventional pressing process. The unloaded density value increases as the load is increased to 1000 MPA (145 ksi) as shown in Figure 1. The yield point of the shaping tooling is indicated as dotted line at 2000 MPa (290 ksi). In the DMC process, due to its dynamic nature, the DMC unloaded green density is 1.85-1.95 g/cc. The crush curve is generated by static loading and does not factor in the dynamic loading component produced by the DMC process. Thus there is a disconnect between the crush curve modeling prediction of 1.55 g/cc at tool yield point and the actual DMC results. This higher DMC green density offers several benefits such as shorter sintering cycle, controlled shrinkage for net shape and green machining for cost savings.

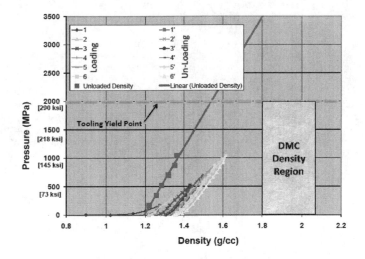

Figure 1. Compressibility curves of 490 NDP measured at Georgia Tech

POWDER COMPACTION

The 490 NDP powders were compacted using different compaction energies to obtain the best possible green densities. The green density of the DMC compacted part increased with compaction energy reaching a plateau asymptotically as shown in Figure 2.

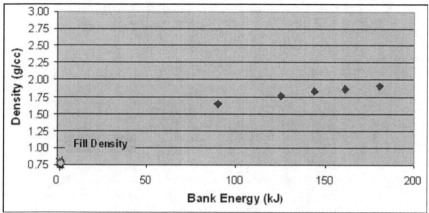

Figure 2. Green density change with compaction energy

The compacted powder part was subjected to a short cure to impart sufficient strength for extraction from tooling and green machining. The green density of the part processed under various test conditions was measured and found to be uniform across the single sample and in all the three tiles pressed in a single stroke. The curing time was optimized to obtain high green density with sufficient strength to handle green machining of the part. Figure 3 shows the uniformity of green and sinter density in all three parts compacted in a cycle. In each cycle, the green density was measured in one sample while the other two were sintered and used for sinter density measurements.

Shrinkage of the samples (of high green density >1.85 g/cc) after sintering was isotropic in all three directions (length x width x height) and was consistently predictable with an average of 15% in all three directions. Figure 4 shows the shrinkage in three directions of 20 tiles in two data sets (compacted at two different energy levels) and their consistency. The shrinkage is uniform with tight variation in samples compacted at higher energy. The DMC shrinkage is thus lower than those obtained in conventional pressing (about 19-20%). In conventionally pressed samples, depending on the density gradients, the shrinkage is anisotropic.

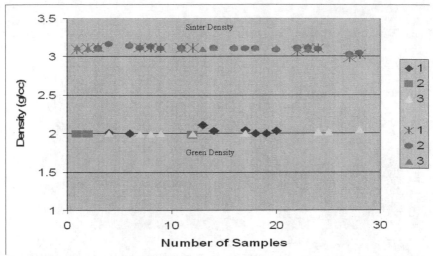

Figure 3. Green density uniformity in each sample and across all three samples (denoted 1, 2, 3) produced in a single compaction cycle. Two runs at different compaction energies were utilized. The graph translates to uniformity in the sinter density.

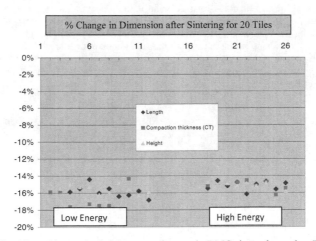

Figure 4. Predictable and isotropic shrinkage consistency in DMC sintered samples. The green density is in the range of 1.85-1.95 g/cc.

SINTERING OF DMC PRESSED SAMPLES AND THEIR PROPERTIES

DMC samples were pressure-less sintered (PS) using the same temperature-time cycle as the conventionally pressed samples. Few variations in sintering times were attempted; however, systematic study of the time and temperature changes could not be performed due to resource and time limitations. Based on test conditions sintered density range of 3.12-3.16 g/cc was obtained in the SiC tiles. The uniformity and sample to sample reproducibility of density were remarkable. The sintered tiles were NDE tested using x-ray computer tomography and were found to be free of cracks.

The sintered samples were measured for their density, hardness (Knoop) and modulus. The density was measured using Archimedes method according to ASTM standard. Figure 5 summarizes the density, hardness, modulus and Poison's ratio of DMC, hot pressed and conventionally processed SiC samples. The conventional samples with larger grain size (right of Figure 7) showed lower Knoop hardness value of 2471±76 at 300 gms load with 15 minutes dwell time relative to DMC samples with finer grain size (left of Figure 7) that showed higher hardness of 2610±49.

After pressure-less sintering the DMC samples showed finer grain size than conventionally pressed samples. When compacted with high energy the DMC green density (> 1.8 g/ cc), and the pressure-less sintered microstructure was similar to that of hot pressed material as shown in Figure 6. For high DMC compaction energy, the starting powders of SiC may be fragmenting into finer powders during compaction. When sintered, such finer powders appear to retain the grain size (Fig. 6). Figure 7 shows samples compacted at lower compaction energy with green density less than 1.8 g/cc. Also shown are results of the DMC pressed sample (green density 1.6 g/cc) sintered side by side with conventionally pressed samples (1.6 g/cc green density) using pressure-less sintering. After conventional pressing the sintered sample showed an average grain size of 16.86 μm (averaged over 60 grains) and sigma value of 1.224. The DMC samples showed about half the grain size when measured over 104 grains.

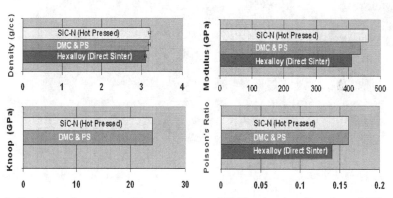

Figure 5. Density, hardness and modulus comparison of DMC and pressure-less sintered (PS) samples with hot pressed and conventionally pressed samples.

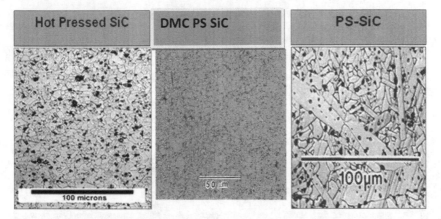

Figure 6. Comparison of microstructure of Hot pressed, DMC-PS and Conventional PS-SiC samples

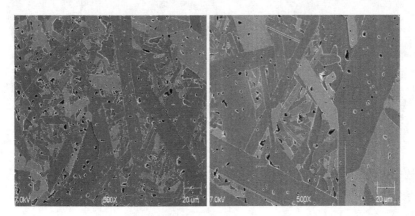

Figure 7. DMC pressed sample (left) of green density of 1.6 g/cc and conventionally pressed sample (right) of green density of 1.6 g/cc – both samples were pressure-less sintered under the same sintering conditions.

SUMMARY

We have developed a new dynamic ceramic processing method that yields 37% higher green compact density than conventional pressing. The shrinkage of high green density compacts is uniform and isotropic in all three sample directions. Machining of high green density samples to final shape was conducted before sintering to minimize sintered machining/grinding. Higher green compact density also offers the potential for sintering for shorter time/or lower temperature. Microstructure and properties of DMC processed ceramics are comparable to hot pressed material. The DMC processed samples yielded higher production rates relative to hot pressed material. A high pressing throughput rate greater than 1350 lbs/day was achieved using the DMC method with an estimated cost below $35 per 4" x 4" x 0.75" tile (< $25/lb) once the system is scaled for production. The ballistic performance of the tiles remains to be tested.

REFERENCES

1. Chelluri, Bhanumathi; and Barber, John; "Structure and Method for compaction of Powder-Like Materials", U.S Patent No. 5,405,574., 5,611,139, 5,639,797, 5,611,230
2. Chelluri Bhanu; "*Full Density Net Shape Powder Consolidation Using Dynamic Magnetic Pulse Pressures*", Journal of Metals **51**, p.35-37 July 1999.

DEVELOPMENT OF BIOMORPHIC SiSiC- AND C/SiSiC-MATERIALS FOR LIGHTWEIGHT ARMOR

Bernhard Heidenreich, Matteo Crippa, Heinz Voggenreiter
DLR – German Aerospace Center
Institute of Structures and Design
Pfaffenwaldring 38-40
D-70569 Stuttgart

Elmar Straßburger
Fraunhofer Gesellschaft e.V.
EMI – Ernst Mach Institute
Am Christianswuhr 2
D-79400 Kandern

Heiner Gedon, Marco Nordmann
WIWEB-Wehrwissenschaftliches Institut
für Werk- und Betriebsstoffe
Institutsweg 1
D-85435 Erding

ABSTRACT

Armour systems based on ceramic materials offer excellent protection against armour piercing ammunition at significantly lower areal densities compared to conventional hard armour steel or aluminium. Their main disadvantages are the high costs of the commonly used monolithic ceramic tiles and the inadequate multi-hit performance.

In the work presented, the manufacturing process, the physical properties and the ballistic behaviour of newly developed biomorphic SiSiC and C/SiSiC (Carbon fibre reinforced SiSiC) ceramics, based on wood powder and activated carbon, have been studied. All samples have been manufactured via the cost effective liquid silicon infiltration process (LSI) developed by DLR. The ballistic performance was studied on sample plates 100 mm x 100 mm with armour piercing ammunition (7.62 x 51 mm AP). The novel SiSiC materials have shown good single hit properties, whereas an increased damage tolerance was obtained by C/SiSiC materials.

INTRODUCTION

For ballistic protection against small calibre armour piercing ammunition, impacting with high projectile velocities of up to 900 m/s, ceramic armour systems are state of the art. Compared to armour steel or aluminium, ceramic materials offer significantly lower weight and therefore are mainly used for personal and vehicle armour, especially in aircrafts and helicopters. Most commonly used materials are monolithic ceramics such as Al_2O_3, SiC and B_4C. The main drawbacks of ceramic armour are their low performance against multiple hits, due to high brittleness and low fracture strength, and their higher cost compared to metallic armour materials.

Since many years, Al_2O_3 armour materials, manufactured via slip casting/sintering methods are widely used due its low price (10 to about 20 US$/kg) compared to SiC and B_4C. SiC is used as reaction-bonded SiC (RBSC or SiSiC; > 35 US$/kg), sintered SSiC or liquid phase sintered LPSSC (> 80 US$/kg) and hot pressed SiC (HPSiC). For boron carbide, mainly hot pressing is used, leading to costs of at least 150 US$/kg. For the manufacture of reaction bonded SiC, preforms based on coarse SiC particles and carbon are infiltrated with molten silicon, embedding the SiC particles in a dense SiSiC matrix. Typical Si contents are in the range of 10 weight-%. For sintered SiC, very fine-grained SiC powder (< 1µm) with sinter additives, like B_4C or C, is used as raw material. After sintering at temperatures of about 2000 °C a residual porosity < 1% can be achieved. For HPSiC an additional pressure of about 2000 bar is applied during sintering, leading to materials almost free of pores. LPSSiC is also a highly dense SiC, based on very fine grained SiC particles and secondary phases of oxidic or oxinitridic ceramics, leading to a higher fracture strength and toughness compared to SiSiC and SSiC. [1]

At DLR, novel biomorphic SiSiC materials have been developed and tested successfully for armour applications in the last years [2-4]. Thereby wood based preforms are converted to SiSiC materials using the so called liquid silicon infiltration process (LSI), a robust and cost efficient manufacturing process, originally developed at DLR for ceramic matrix composites (CMC), especially for carbon fibre reinforced C/C-SiC materials, typically used for thermal protection systems of spacecraft [5] or high performance brake discs [6].

In the first step of the LSI process a green body or preform, based on low cost raw materials e.g. wood fibres and phenolic resin is manufactured via warm pressing. After pyrolysing the preform at temperatures of up to 1650 °C in inert atmosphere in the second step, the resulting, porous C-preform, is siliconized at temperatures above 1450 °C in vacuum. Thereby, molten silicon is infiltrated in the open porosity of the C-preform by capillary forces only and reacts with the carbon, forming SiSiC [7]. The final composition of the ceramic material, i.e. the content of SiC, Si and C, is heavily influenced by the porosity and microstructure of the C-preform and can be varied in a wide range by using tailored green bodies. Due to practically no change in geometry during siliconization and reproducible contraction rates during pyrolysis as well as due to a unique in situ joining technology, even large and complex shaped parts can be manufactured in a cost effective, near net shape technique (Fig. 1).

Fig.1: Examples of biomorphic SiSiC components, manufactured in near net shape technique. Left: Feasibility study of a complex structure (300 x 300 x 50 mm³); Right: In situ joined crucible (340 x 120 x 70 mm³, t = 10 mm)

In ballistic impact tests, biomorphic SiSiC materials demonstrated a high potential for the use in lightweight ceramic armour systems [2-4]. However, biomorphic SiSiC materials based on commercially available, medium density fibre boards (MDF) showed some disadvantages, like a non satisfying multiple hit performance, comparable to conventional armour ceramics. In order to increase the multiple hit resistance, novel materials based on the combination of biomorphic SiSiC and C/C-SiC materials have been developed and tested. Thereby C-fibres were integrated in biomorphic SiSiC to increase ductility and damage tolerance. To increase the ballistic limit velocity, the increase of density and SiC content by varying the composition of the wooden preform, was in the focus.

Another aspect in this work was the feasibility of large, complex shaped geometries like curved armour structures. Thereby the contraction of the wooden preform during pyrolysis is critical. High contraction rates may lead to distortion, spring back effects and even cracks in the C-preform during pyrolysis. Therefore new material variants based on thermally stable additives like carbon fibres or particles have been developed and tested.

At DLR, polycrystalline Si, also called solar grade Si is used as a standard material for the manufacture of hot structures based on carbon fibre reinforced SiC, the so called C/C-SiC material, as well as for biomorphic ceramics, up to now. Solar grade Si offers high purity > 99.99 weight- %, which is favourable for materials and structures required for the use at very high temperatures of 1600 °C and well above, e.g. for heat shields of spacecraft or for rocket motor components. Thereby, impurities would lead to corrosion and oxidation processes, limiting the temperature stability of the materials. The drawback of the solar grade Si is its high cost of up to 90 €/kg, due to high process costs and expensive raw materials. Taking into account, that armour materials are used at low temperatures and that biomorphic SiSiC is made of about 80 weight-% of Si, it is obvious, that cheaper types of Si have a high potential for cost reduction. Therefore, low cost metallurgical Si, widely used as alloy for the manufacture of tinplate and corrosion resistant steel as well as for a raw material for the production of silicone, has been investigated for the manufacture of biomorphic SiSiC armour materials. Metallurgical Si is obtained by the reduction of Quartz with wood coal ($SiO_2 + 2C \rightarrow Si + 2\ CO$) at temperatures of about 2000 °C in electric arc furnaces, leading to a purity of up to 99.8 weight-% Si and low cost of about 5 €/kg.

MATERIALS AND MANUFACTURING

In this work, eight different biomorphic SiSiC and C/SiSiC material variations were investigated in total. All of these materials were based on commercially available raw materials or preforms. The material variations can by classified in three different types of green bodies, which were either commercially available medium dense fibre boards (MDF), wood based composites (WBC) or activated carbon based composites (ACBC), the latter both developed at DLR. In order to increase material ductility and multiple hit performance, short carbon fibres were integrated in three material variants based on WBC (WBC-A1) and ACBC (ACBC-D2, ACBC-E2). To reduce material costs, the high priced solar silicon, was replaced by low cost metallurgical silicon (HQ) in two material variations (MDF-HQ, WBC-B1-HQ).

Manufacture of ceramic tiles

MDF boards are widely used in the furniture industry and are made by pressing of fine fibres of needle wood with binders based on formaldehyde or phenolic resins in a mass production process (worldwide production 2007: 56 x 10^6 m³ = ~35 x 10^6 tons). Thereby very large panels of typically 1220 x 2440 mm² (4" x 8") up to 2.8 m x 6.5 m are manufactured at low costs of about 1.3 €/kg.

For the WBC green bodies, milled wood powder (grain size < 30 µm) and powdery phenolic resin (grain size < 15 µm) were mixed in a dry process using an Eirich mixer (model RV 02 E). After filling in the press mass in a mould, the compound was uniaxially densified in a press and cured at a maximum temperature of 185 °C, leading to sample plates 335 mm x 335 mm x 15 mm (WBC-A1). For the ACBC preforms, activated carbon (grain size 95 µm) was used instead of the wood powder.

The material variants WBC-B1 and ACBC-E2 were based on the materials WBC-A1 and WBC-C2, respectively, but were characterized by adding milled, pitch based graphite fibres (fibre length l_F = 0.37 mm), CF 0.37, into the press mass, prior to the mixing process. To investigate the influence of fibre length to the material behaviour, cut carbon fibres (Tenax HTA, 3K) with a significantly higher length (l_F = 40 mm), CF 40, were used for the material variation ACBC D2. The composition of the different press masses are shown in detail in figure 2.

In order to achieve a near net shape manufacturing of the ceramic armour tiles, the MDF panel was cut to small sample plates (130 mm x 130 mm x 21 mm) using a circular saw. To remove the moisture content of the MDF preforms as well as the water content, developed by the curing process of the phenolic resin in the WBC and ACBC preforms, all the sample plates were dried for 72 h (MDF) and 24 hours (WBC, ACBC) at 110 °C in air before pyrolysis.

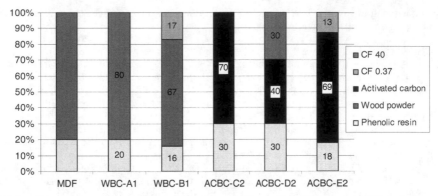

Fig. 2: Composition of the press masses for the manufacture of the WBC and ACBC preforms.

After pyrolysing the preform plates in N_2 atmosphere at $T_{max.}$ = 1550 °C, the WBC and ACBC based carbon preforms were cut to sample plates (100 mm x 100 mm) by diamond saw cutting.

In the last process step, all the carbon preforms were siliconized in vacuum, at $T_{max.}$ = 1550 °C. Thereby, granulated silicon was melted and infiltrated in the porous C preforms by capillary forces only. After cooling down, the biomorphic SiSiC tiles were ground to a final thickness of 9 mm with diamond coated grinding disc.

MDF-HQ and WBC-B1-HQ were based on identical preforms and process parameters as for MDF and WBC-B1, but, for siliconization, the solar grade Si was replaced by a high quality metallurgical grade silicon (Silgrain® HQ, 99,7 % Si), supplied by Elkem AS.

SAMPLE PREPARATION

Samples for determination of mechanical properties

For the determination of the mechanical properties, test specimen (45 mm x 5 mm x 3 mm) for 4 point bending tests were machined out of the sample tiles (100 mm x 100 mm x 10 mm). In a first step the tiles we ground to a thickness of 3 mm, using diamond coated grinding wheel and subsequently cut via diamond coated cutting wheel or by wire-cut EDM (Electrical Discharge Machining).

Single Edge V-Notch Beam (SEVNB) specimens were used for the determination of fracture toughness K_{IC} at Fraunhofer IWM, Freiburg. Thereby a notch was machined in the middle of the specimen (48 x 5 x 4 mm³) using a diamond coated cutting wheel (0.2 mm width). The resulting rectangular notch was prolonged to a V-notch by machine grinding using a razor blade and diamond paste. The final V-notch was characterized by a notch radius below 20 µm and a total depth of 1 mm to 1.4 mm.

Ballistic targets

For ballistic testing, 56 target samples were manufactured in total, 7 targets from each ceramic material variation. The samples were built up as two layer targets consisting of ceramic tiles (100 mm x 100 mm) bonded onto an aramid fibre based backing material, widely used for ballistic protection. The backing panels were build up by 20 layers Twaron T 750 prepreg, with a density of 1.13 g/cm³, resulting in a thickness of 10 mm and an areal density of 11.3 kg/m². The ceramic tiles were bonded to the aramid backing via autoclave technique at elevated temperature and pressure using a polyurethane adhesive (Sikaflex) and an earlier developed procedure at Tejin Twaron, Wuppertal. Subsequently, the ceramic tiles were wrapped with one layer of fibreglass fabric preimpregnated with epoxy resin, which was cured during the autoclave process. Due to the different densities and the constant wall thickness of the ceramic tiles, the final areal density of the sample targets varied between 34.75 and 38.31 kg/m² (table I and II).

Table I: Overview of sample targets based on not fibre reinforced, biomorphic SiSiC tiles.

Target		MDF [1]	MDF	WBC-A1 [1]	WBC-A1	ACBC-C2	MDF-HQ
Tile thickness	[mm]	8	9	8	9	9	9
Backing thickness	[mm]	10	10	10	10	10	10
Number of aramid layers in backing	[-]	20	20	20	20	20	20
Areal density	[kg/m²]	34	36.71	36.4	38.31	37,76	36.76
Ballistic limit velocity	[m/s]	775	930	835	1024	1056	872
Number of targets	[-]	4	7	4	7	7	7

[1] Sample targets from previous investigations [4]

Table II: Overview of targets based on C-fibre reinforced biomorphic C/SiSiC and C/C-SiC tiles.

Target		C/C-SiC XB [1]	C/C-SiC SF [1]	WBC-B1	WBC-B1-HQ	ACBC-D2	ACBC-E2
Tile thickness	[mm]	10	10	9	9	9	9
Fibre type	[-]	HTA	HTA	graphite	graphite	HTA	graphite
Fibre length	[mm]	endless	40	0.37	0.37	40	0.37
Backing thickness	[mm]	10	10	10	10	10	10
Number of aramid layers in backing	[-]	20	20	20	20	20	20
Areal density	[kg/m²]	31.8	31.9	37.9	37.2	34.75	38.26
Ballistic limit velocity	[m/s]	594	457	959	944	765	1055
Number of targets	[-]	4	4	7	7	7	7

[1] Sample targets from prior investigations [2, 4]

TEST PROCEDURE

Mechanical properties
Flexural strength and Young´s modulus were determined in four point bending test in accordance with DIN EN 843-1 and -2. Young's modulus was calculated from elastic deformation in a force margin from 50 N to 200 N. Flexural strength results from maximum bending moment and section modulus. Peripheral strain was calculated by using a trapeze-shaped deformation model. Then the strain is given by $\varepsilon = \dfrac{\Delta l_R}{l_A} = 2 \cdot d \Big/ l_A \cdot \tan\left\{\dfrac{1}{2}\left[\arctan\left(\dfrac{l_a}{h} + \dfrac{\pi}{2}\right)\right]\right\}$, where l_A is the distance of the lower bearing, l_a the difference of lower and upper bearing, Δl_R the peripheral strain, h the vertical bending and d the sample thickness. Fracture toughness (critical stress intensity factor) K_{IC} was also determined in four point bending, using the SEVNB-method, according to ASTM STP 1409. K_{IC} was calculated using the formula $K_{IC} = F_{max} / b \cdot \sqrt{d} \cdot Y$ where Y correlates crack length with

geometry, b= sample width, d = sample thickness and F_{max}= maximum load. Density and open porosity were determined from the tiles used for ballistic tests as well as from the bending samples using the water immersion method based on Archimedes law. Hardness was examined according to Vickers and Knoop with a load of 1 kg referring to DIN EN 843-4. To get a representative average hardness value, for each material three valid tests out of several measurements, randomly distributed over a polished sample, were taken into account. Crack surfaces and cross section cuts were examined by scanning electron microscopy (SEM) using secondary and back scattered electron mode (SEI, BSE). Phase contents were determined by a quantitative SEM evaluation based on grey-scale analysis.

Ballistic tests

The ballistic performance of the different ceramic materials was tested with 7.62 mm x 51 AP armour piercing ammunition with a total projectile mass of 9.5 g and a steel core mass of 3.7 g. In order to determine the ballistic limit velocity v_{BL} of each material, the impact velocity (v_P) of the projectile was varied between 830 and 1115 m/s. The samples were fixed by clamping the aramid backing plates between two steel frames. Each sample was tested with a single hit. The ballistic limit velocities were determined using the Lambert -Jonas approach: $v_R = [K(v_P^2 - v_{BL}^2)]^{1/2}$, with v_R = residual velocity after perforation [8]. In this approach only the balance of the kinetic energy is considered. A curve of the Lambert-Jonas type delivers a good approximation of the experimental data. The parameter K and the ballistic limit velocities v_{BL} were determined by a least squares fit of the experimental data.

RESULTS

Contraction behaviour during pyrolysis

Due to the volume contraction of both, the wood powder and the binder, i. e. the phenolic resin, the MDF and WBC-A1 preforms showed the highest mass losses (69 % and 65 %) and volume contraction (66 % and 63 %) during pyrolysis (Fig. 3). By adding thermally stable C-fibres and activated C-powder, a minimum mass loss of 20 % and volume contraction of 9 % could be achieved. The mass losses where related directly to the content of phenolic resin and wood powder. The higher the content of thermally stable additives the lower the mass loss. In contrast, the volume contraction, especially in the in plane direction, was not only determined by the C-content itself, but was also influenced heavily by the type and geometry of the C-additives (Fig. 4). Adding C-fibres, the in plane contraction of the preform plates could be reduced dramatically from 23 % and 25 % for MDF and WBC-A1, which were not fibre reinforced, to about 5 % by adding very short graphite fibres (WBC-B1, ACBC-E2) and lead to practically no contraction (0.1 %) using 40 mm long HTA fibres (ACBC-D2). In contrast, the contraction in transverse direction was almost linear to the C-content and only a minor effect of the fibres could be observed. This behaviour was explained by the uniaxial pressing during preform manufacturing, leading to a fibre orientation mainly in the in plane direction, as well as by the restraint to contraction parallel to the C-fibres, a well known effect from the manufacture of C/C-SiC [9].

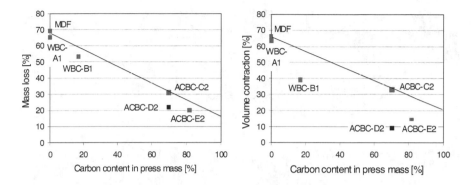

Fig. 3: Mass losses and volume contraction during pyrolysis in dependence of the carbon content (C-fibres, activated carbon) of the press masses.

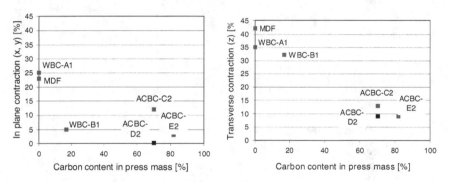

Fig. 4: Transverse and in plane contraction of the preforms during pyrolysis in dependence of the carbon content (C-fibres, activated carbon) of the press masses.

Microstructure

Main phases of all investigated materials were β-SiC and Si. The amount of residual C was lower than 6 % for all materials. Si was distributed rather uniformly between the SiC grains which showed an average grain size of 10 to 20 µm. However, it is important to mention that in some cases big agglomerates (up to 100 µm) as well as linked network structures of SiC were observed, where grain boundaries were not clear to determine (Fig. 5).

The biomorphic SiSiC materials were relatively dense with open porosities in the range of 0.00 to 0.24 %. Density varied between 2.6 g/cm³ and 3.0 g/cm³ (fig. 6). Thereby, as expected, the material variant ACBC-D2 with the high volume content of long C-fibres ($l_F = 40$ mm; $\phi_F = 30$ vol.-%) showed the lowest density, due to the high amount of residual Si as well as C from fibres and matrix not converted to SiC during siliconization (Fig. 8). As a result of the high Si-content, the density of MDF based materials was limited to 2.8 g/cm³. Highest densities were obtained by the materials based on wood powder or on activated coal, which was explained by a favourable porous microstructure of the carbon preform after pyrolysis, leading to a high conversion rate of C to SiC ($\phi_{SiC} > 75$ vol.-%) after siliconization. Adding graphite fibres did not influence the density, due to the fact that the fibres were converted almost totally to SiC during siliconization.

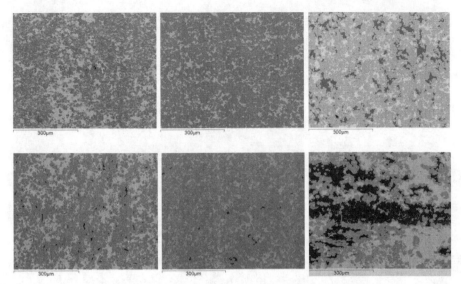

Fig. 5: SEM images (200 x, cross section) of biomorphic SiSiC consisting of SiC (dark grey), C (black) and Si (light grey). Top: Not C-fibre reinforced SiSiC based on MDF (left), WBC-A1 (middle), and ACBC-C2 (right) preforms. Bottom: Biomorphic SiSiC derived by C-fibre reinforced preforms. The graphite fibres (l_{fibre} = 370 μm) added to WBC-B1 (left) and ACBC-E2 (middle) are converted to SiC and not visible anymore. Right: ACBC-D2 based on cut HTA-fibres (l_{fibre} = 40 mm) with C/C-bundles (black), partially not converted to SiC, clearly visible.

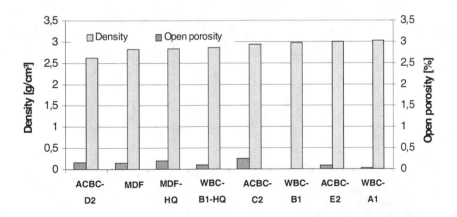

Fig. 6: Density and open porosity of the biomorphic SiSiC materials, determined on tiles for ballistic impact testing.

Mechanical properties

The mechanical properties of the biomorphic ceramics are mainly defined by their phase composition and microstructure. Thereby, a high content of silicon, hence a low density, caused low values for hardness, flexural strength and Young's modulus but relatively high fracture toughness K_{IC}. Regarding the WBC and ACBC materials with a density of $3g/cm^3$, the differences in mechanical properties are marginal. High values of Young's modulus (360 – 370 GPa) and flexural strength (170 – 190 MPa) could be obtained, comparable to conventional RBSC ceramics (table III). The fibre reinforced materials with a high density, i.e. the materials with the short, graphite fibres added, did not show a quasi-ductile behaviour or significantly higher values of K_{IC}, due to the conversion of the fibres to SiC. In contrast, the relatively long HTA C-fibre bundles lead to more quasi-ductile fracture behaviour in ACBC-D2 (Fig. 7). Thereby the C-fibres partially survived the siliconization process and fibre pullout could be observed (Fig.8). Determination of K_{IC} was not possible, because crack propagation started outside the angular point of the V-notch. Young's modulus, strength and hardness were significantly lower compared to the not fibre reinforced materials and compared to the materials, in which the fibres were converted almost totally to SiC.

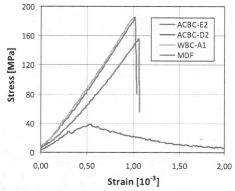

Fig. 7: Stress-strain curves of selected specimen, showing brittle fracture behaviour for not fibre reinforced WBC-A1 and ACBC-E2 (C-fibres converted to SiC), typical for monolithic ceramics, and a quasiductile behaviour of fibre reinforced ACBC-D2, resulting from fibre pullout effects.

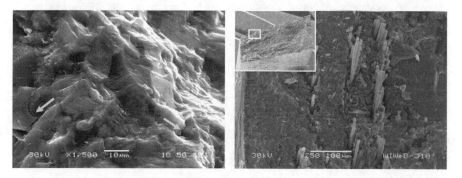

Fig. 8: SEM images of fracture surfaces of ACBC-E2 (left), showing intergranular crack propagation and secondary cracks between SiC and Si as well as intragranular cracks in a residual carbon grain. Right: ACBC-D2 with partial fibre pull out, clearly visible.

Table III: Physical properties of the investigated biomorphic SiSiC materials.

	MDF	WBC-A1	ACBC-C2	WBC-B1	ACBC-D2	ACBC-E2	MDF-HQ	WBC-B1-HQ
Fibre type [-]	-	-	-	graphite	HTA	graphite	-	graphite
Fibre content [vol.-%]	0	0	0	17	30	13	0	17
Density [g/cm³]	2.81	3.01	3.01	2.93	2.66	3.04	2.84	2.87
Flexural strength [MPa]	160	190	170	180	40	190	170	210
Young's modulus [GPa]	310	360	370	320	230	370	300	340
K_{IC} [MPa×m$^{1/2}$]	2.33	2.71	2.95	2.,9	-	2.74	2.17	2.86
Vickers hardness HV1 [kg/mm²]	1400	1950	2250	2150	1200	2200	1350	1400
Knoop hardness HK1 [kg/mm²]	1250	1650	1700	1650	1000	1750	1350	1650

Ballistic properties

Fig. 9 is summarizing the results of the ballistic tests, showing the residual projectile velocity v_R after impact of the target as a function of the impact or projectile velocity v_p. The intersection of the Lambert-Jonas curve with the abscissa ($v_R = 0$ m/s) is marking the ballistic limit velocity v_{BL}.

The biomorphic SiSiC derived from MDF could stop the projectiles with v_p of up to 902 m/s. In two test with $v_p = 1013$ m/s and 1049 m/s, the projectile perforated the armour system ($v_R = 529$ m/s and 758 m/s). In another two tests ($v_p = 974$ m/s), the armour system was perforated, too, but v_R was not detected. Ballistic limit velocity was calculated to $v_{BL} = 930$ m/s. With WBC-A1 the projectiles could be stopped in five tests with $v_p = 839$ m/s up to $v_p = 1048$ m/s. The ballistic limit velocity was calculated to $v_{BL} = 1024$ m/s. From all the biomorphic SiSiC materials, not reinforced with C-fibres, the highest ballistic limit velocity of $v_{BL} = 1056$ m/s was obtained with ACBC-C2. Thereby, the projectiles could by stopped in 5 tests with $v_p = 842$ m/s up to $v_p = 1052$ m/s. In two tests at $v_p = 1041$ m/s and 1104 m/s, the targets were perforated ($v_R = 435$ m/s and 645 m/s).

Biomorphic SiSiC materials, based on C-fibre reinforcement showed significantly different ballistic limit velocities. The targets based on ACBC-D2 ($\phi_F = 40$ vol.-%; $l_F = 40$ mm) showed a relatively low ballistic limit velocity with $v_{BL} = 765$ m/s. In four tests with $v_p = 837$ m/s up to $v_p = 1092$ m/s the projectiles could not be stopped. The targets were perforated in all tests. However, in three tests the residual velocities could not be determined. In contrast, the materials based on the graphite fibres showed much higher ballistic limit velocities of $v_{BL} = 959$ m/s (WBC-B1) and $v_{BL} = 1055$ m/s (ACBC-E2). Thereby, WBC-B1 could stop the projectiles in 5 tests with $v_p = 839$ m/s to 1092 m/s. The ACBC-E2 targets could stop the projectiles in two tests at $v_p = 1055$ m/s and 1057 m/s. In five tests with $v_p = 1058$ m/s to 1111 m/s the projectiles completely penetrated with v_R in the range from 393 m/s to 600 m/s.

Biomorphic SiSiC tiles based on low cost, metallurgical Si (HQ) showed ballistic limit velocities similar to their counterparts based on solar grade Si. The targets based on MDF-HQ could stop the projectiles at $v_p = 913$ m/s and 917 m/s. In five tests with $v_p = 903$ m/s to 1023 m/s the armour systems were perforated with $v_R = 268$ m/s to 653 m/s, resulting in $v_{BL} = 872$ m/s, slightly lower compared to MDF ($v_{BL} = 930$ m/s). The ballistic limit velocity of WBC-B1-HQ was almost identical with WBC-B1 ($v_{BL, WBC-B1} = 959$ m/s; $v_{BL, WBC-B1-HQ} = 944$ m/s).

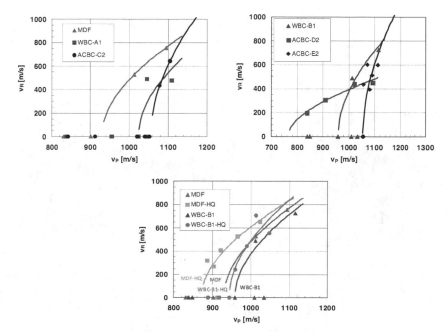

Fig. 9: Residual velocity as a function of the projectile impact velocity. Comparison of not C-fibre reinforced (top left) and C-fibre reinforced (top right) materials. Bottom: Comparison of biomorphic SiSiC siliconized with solar grade and metallurgical Si (HQ).

In every case where the projectile was stopped, the hardened steel core was broken, but the ceramic tiles were strongly fragmented. Due to the deformation of the aramid backing, most of the ceramic fragments were separated from the backing. However, comparing ballistic tests with similar projectile and residual velocity, the fibre reinforced ACBC-D2 ($l_F = 40$ mm) tiles showed a more favourable fracture behaviour, compared to the not fibre reinforced materials (Fig. 10). As expected, the fracture behaviour of the tiles based on the short (370 μm) fibre reinforced preforms showed no significant improvements, compared to not fibre reinforced materials, due to the conversion of the C-fibres to SiC during siliconization.

At first glance, the ballistic limit velocity seems to be almost proportional to the areal density of the armour system or, hence the weight of the backing is the same in all targets, to the areal density of the ceramic tile, respectively. The higher the areal density of the ceramic tile and therefore the higher its density and/or wall thickness, the higher the ballistic limit velocity (Fig. 11). However, the ballistic performance may be ranked more sound by the specific ballistic limit velocity ($v_{BL,spec.}$), i.e. the ballistic limit velocity divided by the areal density (Fig. 12). Thereby it could be shown, that the highest ballistic performance could be obtained by ACBC-C2 and ACBC-E2 with $v_{BL, spec} = 27,6$ and 28 (m/s)/(kg/m²), $v_{BL} = 1055$ m/s and areal densities of 37.8 and 38,3 kg/m², which was explained by the high SiC-content and density of the ceramic materials.

MDF offered the same specific v_{BL} as WBC-B1, due to its slightly lower ballistic limit velocity but significantly lower areal density (MDF: $v_{BL} = 930$ m/s; m = 36.7 kg/m², WBC-B1: $v_{BL} = 959$ m/s; m = 37.9 kg/m²). Compared to WBC-A1 the ballistic performance of MDF was even higher, due to the significantly higher ballistic limit velocity at similar areal density (MDF (t = 9mm): v_{BL} = 930 m/s; m = 36.7 kg/m², WBC-A1 (t = 8 mm): $v_{BL} = 835$ m/s; m = 36,4 kg/m²). However, for

WBC-A1 a lower wall thickness is sufficient for proper protection, which can be advantageous applications with limited space available.

The C/C-SiC materials XB and SF showed the lowest specific ballistic limit velocity, due to their high C-content (fibres and matrix). In contrast, despite its residual C-fibre content, ACBC-D2 offered a significantly higher specific ballistic limit velocity compared to XB and SF, similar to the specific ballistic limit velocity of MDF and WBC-A1, determined on 8 mm tiles.

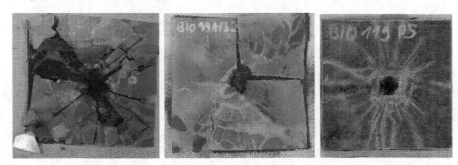

Fig. 10: Comparison of fracture behaviour of biomorphic SiSiC tiles with and without C-fibre reinforcement after impact. Left an middle: Not fibre reinforced MDF (v_P = 1094 m/s; v_R = 758 m/s) and WBC-A1 (v_P = 1046 m/s; v_R = 490 m /s) showing heavy fragmentation. Right: ACBC-D2 with C-fibre reinforcement (l_F = 40 mm; v_P = 1022 m/s, v_R = 441 m/s) showing significantly less fragmentation.

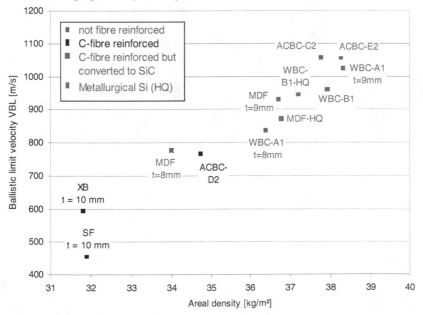

Fig. 11: Overview of ballistic limit velocity in dependence of the total areal density of the armour system (9 mm ceramic tile and aramid backing). Results from former investigations [2, 4], based on ceramic tiles with a thickness of 8 and 10 mm are integrated for comparison.

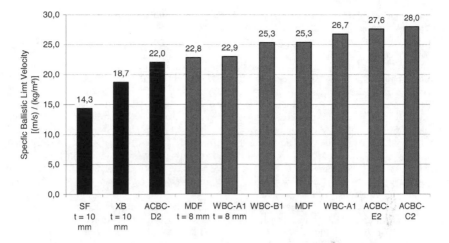

Fig. 12: Overview of specific ballistic limit velocities of the tested armour systems (9 mm ceramic tile and aramid backing). Results from former investigations [2, 4], based on ceramic tiles with a thickness of 8 mm and 10 mm are integrated for comparison.

SUMMARY

In this work, 8 different, biomorphic SiSiC materials for lightweight armour systems have been developed and tested against 7.62 x 51mm AP ammunition in single shots. Ceramic armour tiles (100 x 100 x 9 mm³) were manufactured in near net shape technology via the cost efficient LSI process, using commercially available low cost raw materials, i.e. phenolic resins, wood powder, activated carbon, milled or cut C-fibres and metallurgical Si, as well as MDF preforms.

The material development was focused on biomorphic SiSiC-materials with high SiC content, to obtain high mechanical properties and ballistic performance on the one hand as well as on novel, fibre reinforced ceramics, combining high ballistic performance with fracture toughness and multiple hit performance on the other hand. Another main emphasis was the further development of tailored green bodies or preforms, characterized by low changes in geometry during the manufacturing process, enabling a cost efficient and reproducible manufacture of ceramic armour structures in near net shape technique, especially for complex shaped geometries.

Due to favourable microstructures after pyrolysis, almost dense SiSiC materials characterized by low open porosities < 0.24% high densities of about 3 g/cm³ and calculated SiC contents of about 75 vol.-% and above could be achieved by converting preforms based on wood powder (WBC-A1) and activated carbon (ACBC-C2) as well as on their C-fibre reinforced counterparts (WBC-B1, ACBC-E2). For the latter, milled graphite fibres were used, which were converted almost totally to SiC during siliconization. Compared to biomorphic SiSiC based on MDF, the SiC-content and therefore the density could be increased by about 40 % and 7 %, respectively.

Ballistic limit velocities of v_{BL} = 930 to 1056 m/s at low areal densities of 36.7 to 38.3 kg/m² could be obtained by newly developed materials, characterized by high densities of about 3 g/cm³. However, these materials showed brittle fracture behaviour, comparable to conventional monolithic ceramics, thus limiting ballistic protection to single hits. To avoid this general drawback of ceramic armour, a novel material based on the combination of biomorphic SiSiC and quasiductile C/C-SiC has been developed and tested successfully. Thereby a high amount of cut C-fibres (HTA 3K, l_F = 40 mm, ϕ_F = 40 vol.-%) was integrated in a matrix of biomorphic SiSiC.

Compared to not fibre reinforced, biomorphic SiSiC, the resulting ACBC-D2 material showed reduced tile fragmentation after impact in combination with a ballistic limit velocity similar to the MDF based material at comparable areal density, and therefore offers a potential for multiple hit protection.

Due to the use of metallurgical Si, instead of highly pure solar grade Si, the raw material costs for biomorphic SiSiC could be reduced significantly, well below the level of commonly used SiC armour materials, while maintaining the ballistic performance.

The contraction of the green bodies during pyrolysis is most critical for a cost efficient and reproducible manufacturing process based on near net shape technology, especially for the realization of large and curved structures. By adding milled carbon fibres or activated carbon, the volume contraction could be reduced dramatically from 66 % (MDF) to 14 % (ACBC-E2) while maintaining highest SiC-contents in the final ceramic material. Lowest volume contraction of 9 % could be obtained by integrating C-fibres with a length of 40 mm (ACBC-D2). Thereby, the contraction was limited to the thickness direction and practically no change of geometry was observed in the in plane direction.

In a next step, the promising biomorphic SiSiC materials will be ranked by the determination of the minimum areal density, necessary to provide typical ballistic requirements, e.g a ballistic limit velocity of 850 m/s. Future work will be focused on the further improvement of multiple hit performance and the upscaling for the manufacture of large and complex shaped armour structures.

REFERENCES

[1]ETEC Gesellschaft für technische Keramik mbH: SICADUR®/BOCADUR, Company Brochure, Lohmar (2009)

[2]B. Heidenreich, W. Krenkel and B. Lexow, Development of CMC-Materials for lightweight Armour, Proceedings of the 27th Annual Cocoa Beach Conference & Exposition, January 26-31, (2003).

[3]B. Heidenreich, M. Gahr and E. Medvedovski, Biomorphic reaction bonded silicon carbide ceramics for armour applications, Proceedings of the 107th Annual Meeting of The American Ceramic Society, April 10-13, (2005).

[4]B. Heidenreich, M. Gahr, E. Straßburger and E. Lutz, Biomorphic SiSiC-Materials for lightweight Armour, Proceedings of 30th International Conference on Advanced Ceramics & Composites, Cocoa Beach, January 22-27, (2006).

[5]W. Krenkel and H. Hald, Liquid Infiltrated C/SiC - An Alternative Material for Hot Space Structures, ESA/ESTEC Conference on Spacecraft Structures and Mechanical Testing, Noordwijk, The Netherlands, (1988).

[6]W. Krenkel, R. Renz, B. Heidenreich, Lightweight and Wear Resistant CMC Brakes, 7th International Symposium Ceramic for Engines, Goslar, (2000).

[7]M. Gahr, J. Schmidt, A. Hofenauer, O. Treusch, Dense SiSiC ceramics derived from different wood-based composites: Processing, Microstructure and Properties, Proceedings of the 5th International Conference on High Temperature Ceramic Matrix Composites (HTCMC 5), p. 425, The American Ceramic Society (2004).

[8]J.P. Lambert, G.H. Jonas, Ballistic Research Laboratory, Report BRL-R-1852, (1976)

[9]B. Heidenreich, N. Lützenburger, H. Voggenreiter, Carbon Fiber Reinforced Ceramics, in H. Jäger and W. Frohs (Ed.), Reprint from Ullmann`s Encyclopedia of Industrial Chemistry, pp. 91-92, Wiley-VCH, ISBN 978-3-5s7-32445-3, (2009)

INFLUENCE OF IMPURITIES ON STACKING FAULT DYNAMICS IN SIC UNDER EXTERNAL LOADING

Vladislav Domnich and Richard A Haber
Department of Materials Science and Engineering, Rutgers University
Piscataway, NJ, USA

ABSTRACT

The effect of additives (Al, B, N, O) on stacking fault (SF) dynamics in 3C, 2H, 4H, and 6H silicon carbide (SiC) polytypes under external loading is investigated using a combination of an axial next-nearest-neighbor Ising model and single-point energy calculations within the scheme of density-functional theory. Both hydrostatic pressure and pure shear stress are considered. Additives are considered as point defects substituting for either Si or C atoms in the SiC structure. The results of the simulations imply that the (3111) SF in 6H SiC has the highest energy of all possible stacking faults among the SiC polytypes considered. To various degrees, aluminum, nitrogen, and oxygen are found to facilitate SF induced plasticity in SiC. The role of additives in promoting plastic deformation in SiC in a wider pressure range is discussed.

INTRODUCTION

Because the strength of ceramics is crucial for ballistic performance, assessing the behavior of silicon carbide (SiC) under dynamic loading has been the subject of numerous studies. SiC has been shown to maintain its strength up to ~100 GPa, and it appears to exhibit high degree of ductility above the Hugoniot elastic limit (HEL) [1]. The exact mechanism for the large plasticity of SiC at higher impact loads is under debate; however, it may be attributed, in particular, to the slip along the grain boundaries or the formation of large numbers of stacking faults (SF) in the SiC structure.

The SiC stacking fault dynamics at elevated pressures has been theoretically studied by Fanchini [2] using a combination of an axial next-nearest-neighbor Ising (ANNNI) model [3] and single-point energy calculations within the scheme of density-functional theory (DFT) [4,5]. The stacking fault energies (SFE) have been calculated as functions of hydrostatic pressure for 3C, 4H, and 6H SiC polytypes. It has been demonstrated that the formation of SF along the basal plane of 6H SiC is more favorable than formation of SF along any other planes in any other SiC polytype considered. In particular, the onset of SF induced plasticity in 6H SiC was predicted to occur at ~28 GPa, well below a similar onset in the 4H SiC polytype (~67 GPa) or in the 3C SiC polytype (>100 GPa).

Impurities in SiC, composition of the starting powders, core-rim effects will all have effect on polytype distribution, SF dynamics and the resultant plasticity of the material. In this work, we theoretically investigate the effect of additives (Al, B, N, O) on polytype structure and stacking fault energies in SiC at ambient conditions and at elevated pressures. Both hydrostatic pressure and pure shear stress are considered. Additives are considered as point defects substituting for either Si or C atoms in the SiC structure. The potential of additives for enhancing ballistic performance of SiC through promotion of plastic deformation in a wider pressure range is discussed.

EXPERIMENTAL APPROACH

All existing SiC polytypes are tetrahedrally bonded and can be seen as an assembly of corner-sharing tetrahedral. Each tetrahedron consists of four Si (C) atoms at its corners bonded to a C (Si) atom at the centroid of the tetrahedron. There exist several classification schemes for SiC polytypes. [6] In the common Ramsdell notation, each polytype is characterized as nL, where n is the periodicity of the tetrahedra along the c axis, and L is the symmetry of the resulting structure: L = C for cubic, H for hexagonal, and R for rhombohedral symmetry. In the ABC notation, each basal plane, consisting of

Table I. Inequivalent substitutional sites in SiC polytypes. [7]

Ramsdell notation	ABC notation	Jagodzinski notation	No. of inequivalent Si (C) sites	
			cubic-like	hexagonal-like
3C	ABC	k	1	0
2H	AB	h	0	1
4H	ABAC	hk	1	1
6H	ABCACB	hkk	2	1

only C (Si) atoms, is denoted as A, B, or C, depending on the stacking sequences of basal planes along the c direction. The Jagodzinski notation reflects difference in configuration of second neighbors on the same sublattice as the site considered and denotes atoms as k for cubic-type configuration of second neighbors, like that of atom B in the sequence ABC, and h for hexagonal-type configuration of second neighbors, like that of atom B in the sequence ABA. In Hägg notation, (+) is used for AB, BC, CA stacking sequences, and (−) for BA, CB, AC stacking sequences. Finally, Zhdanov notation consists of pairs of numbers in which the first number denotes the number of consecutive plus signs, and the second the number of consecutive minus signs.

Additives in various SiC structures can substitute for Si or C atoms in either k or h sites. The available inequivalent substitutional sites in 3C, 2H, 4H, and 6H SiC polytypes are listed in Table 1 along with their respective symmetry. Boron, nitrogen, oxygen, and aluminum atoms were used as substitutional additives in this study. Substitution for either C or Si atom in both k and h sites was considered.

Extended SF regions can be created in SiC by the motion of partial dislocations, leaving behind a faulted crystal containing an error in the stacking sequence. For symmetry reasons, there exists one non-degenerate SF type in 3C SiC, denoted in Zhdanov notation as (111); one non-degenerate SF type in 2H SiC, denoted as (2); two non-degenerate SF types in 4H SiC, denoted as (31) and (13), respectively; and three non-degenerate SF types in 6H SiC, denoted as (42), (24), and (3111).[8] Different types of stacking faults in SiC polytypes along with their respective symmetry are illustrated in Figure 1.

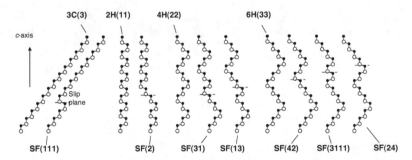

Figure 1. Geometrically distinguishable stacking faults obtained by glide in 3C, 2H, 4H, and 6H SiC in different slip planes (dashed horizontal lines), viewed from the [1120] direction. SiC polytypes and the corresponding stacking faults are classified following Zhdanov notation.

The stacking fault energy can be defined as [6]

$$\gamma = \frac{E_{faulted} - E_{perfect}}{A},$$ (1)

where A is the interface area, $E_{faulted}$ is the total energy of the crystal with a SF, and $E_{perfect}$ is the total energy of the perfect crystal. In the present approach, the total energy refers to the energy of a supercell, and the area $A = a^2 \sqrt{3}/2$ is the area of the supercell in the (0001) basal plane.

The one-dimensional character of the stacking faults in SiC polytypes and the periodic arrangement of the resulting supercells suggests their description with the axial next-nearest-neighbor Ising model formally identical to the one-dimensional Ising model for describing the interactions among coupling spins along one crystallographic axis of the lattice.[3] Restricting the interactions to the next-nearest neighbors, the SF energies in this model can be expressed as (with the subscripts on γ denoting polytype and type of SF, respectively)

$$\gamma_{3C(2)} = \frac{4}{A} \cdot (J_1 + J_2 + J_3),$$

$$\gamma_{2H(111)} = \frac{4}{A} \cdot (-J_1 + J_2 - J_3),$$

$$\gamma_{4H(13)} = \gamma_{4H(31)} = \frac{4}{A} \cdot (-J_2),$$ (2)

$$\gamma_{6H(24)} = \gamma_{6H(42)} = \frac{4}{A} \cdot (-J_3),$$

$$\gamma_{6H(3111)} = \frac{4}{A} \cdot (J_1 - J_2 - J_3),$$

where the "spin interaction" energies J can be obtained in terms of the total energies E_{3C}, E_{2H}, E_{4H}, and E_{6H} per Si-C pair in the perfect crystal using much smaller unit cells:

$$J_1 = \frac{2E_{2H} - E_{3C} + 2E_{4H} - 3E_{6H}}{4},$$

$$J_2 = \frac{-E_{2H} - E_{3C} + 2E_{4H}}{4},$$ (3)

$$J_3 = \frac{-E_{3C} - 2E_{4H} + 3E_{6H}}{4}.$$

The values of E_{3C}, E_{2H}, E_{4H}, and E_{6H} per Si-C pair in the perfect crystal were calculated within the scheme of density-functional theory in the local spin density approximation. [4,5] The Slater functional for the exchange energy [9] and the Vosco, Wilk, and Nusair functional for the correlation energy [10] were used. Commercial software (Gaussian[TM]) [11] with STO-3G Gaussian-type electronic orbitals [12,13] as the basis set was used in our calculations.

When viewed as a hexagonal structure, 3C SiC has 6 atoms in the primitive unit cell, whereas 2H, 4H, and 6H SiC have 4, 8, and 12 atoms per primitive cell, respectively. In order to reduce systematic errors during calculation, the total energy minimizations were performed on supercells containing 24 atoms for all polytypes, i.e. the supercells for 3C, 2H, 4H, and 6H SiC consisting of 4, 6, 3, and 2 primitive unit cells stacked on top of one another, respectively.

For the calculation of stacking fault energies at elevated pressures, the exact lattice parameters as a function of pressure need to be calculated. This is achieved by determining elastic constants in SiC polytypes at specific pressure values. To determine elastic constants of a crystal using total energy and force calculations, a deformation of the supercell is imposed by changing the lattice vectors $\overline{R} = (\overline{a}, \overline{b}, \overline{c})$ of the undisturbed supercell to $\overline{R}' = (\overline{a}', \overline{b}', \overline{c}')$ using a strain matrix [14]

$$
\overline{R} = \overline{R}' \begin{pmatrix} 1 + e_{xx} & \frac{1}{2}e_{xy} & \frac{1}{2}e_{xz} \\ \frac{1}{2}e_{yx} & 1 + e_{yy} & \frac{1}{2}e_{yz} \\ \frac{1}{2}e_{zx} & \frac{1}{2}e_{zy} & 1 + e_{zz} \end{pmatrix}. \tag{4}
$$

The deformation leads to a change in the total energy of the crystal given by

$$
U = \frac{E_{tot} - E_0}{V_0} = \frac{1}{2} \sum_{i=1}^{6} \sum_{j=1}^{6} C_{ij} e_i e_j, \tag{5}
$$

where E_0 is the total energy of the undisturbed lattice, V_0 is the equilibrium volume and the C_{ij} are the elastic constants with $i, j = 1 \ldots 6 = xx, xy, zz, yz, zx, xy$. The elastic constants are determined by calculating E_{tot} as a function of $e_i e_j$ and taking the second energy derivative with respect to e_i and e_j. The hydrostatic pressure is obtained by calculating the change in volume of the supercell with respect to its equilibrium volume at ambient pressure V_0, which are related through the bulk modulus

$$
B = \frac{1}{3}(C_{11} - 2C_{12}). \tag{6}
$$

After finding equilibrium lattice parameters of undeformed crystals by energy minimizations, shear deformation (γ) along the $[01\overline{1}0]$ direction has been applied to each simulation cell. The atomic configurations have been then optimized so that all the forces acting on the atoms are below 0.005 eV/Å; and normal strains of each cell have been adjusted so that normal stress components are negligible. Shear stress τ has been determined for each value of the applied shear deformation using the compliability tensor.

In order to gain the insight on plastic deformation of SiC induced by SF dynamics at elevated pressures, we made assessment of the activation energies and slipping velocities of the stacking faults along the Shockley partials. The dislocation mobility is activated with energy Q as [15]

$$
v_{SP}(\theta, p) \propto \exp\left(\frac{Q - SFE(p)}{k_B T}\right), \tag{7}
$$

where $\theta = 30°$ (90°) is the angle of a particular Shockley partial dislocation, k_B is Boltzmann constant, T is temperature, and p is pressure. Slipping along the stacking faults will occur only when the stacking fault energy at a given pressure SFE(p) is greater than the activation energy Q. The calculations reveal that the activation energy Q is nearly independent of the type of SiC polytype or the pressure. [15] For estimation of the SFE as a function of hydrostatic pressure or shear stress against the activation energy

Q for gliding in the SiC basal plane, the case of 90° Si-core partial dislocation (21.9 meV/atom) [15] was considered as the one that requires minimum activation energy.

RESULTS AND DISCUSSION

Single point calculations of total energy were performed for varying lattice constants a and c near the expected equilibrium position. A total energy surface was then constructed and the energy minimum corresponding to equilibrium lattice constants was determined by polynomial fitting. Once the optimum lattice constants were determined, the energy of the optimized supercell was calculated for each polytype. These total supercell energies, normalized to a single bond, were then used for calculation of spin interaction energies J_1, J_2, J_3.

In a similar fashion, the optimized lattice parameters and the equilibrium supercell total energies were calculated for each polytype with a single B, N, O or Al atom substituting for either Si or C in all available inequivalent h and k sites in a particular polytype. Taking cohesive energy as a criterion for lattice stability, it was observed that the B, N, and O atoms tend to substitute for C, and the Al atom tends to substitute for Si in the SiC lattice of all polytypes considered. This is in agreement with earlier experimental observations reported in literature. [16] Among all possible substitution for the C (Si) atom, the k site was found to be energetically more favorable in all cases. Therefore, only the lattices with B, N, and O in place of C, and Al in place of Si in the k sites were considered in further calculations.

It was noticed that addition of N, O or Al to the SiC structure increases the lattice parameters and the dimensions of the supercell, to the largest extent in case of Al. Inversely, substitution of C for B was leading to a decrease in the lattice parameters and the dimensions of the supercell. Both the lattice parameters a, c and the spin interaction energies J_1, J_2, J_3 calculated in this work showed reasonable correlation with previously published data. [8,17-18]

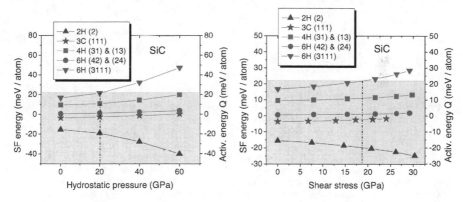

Figure 2. Stacking fault energy as a function of hydrostatic pressure (a) and shear stress (b) in SiC. The onset for plastic transformation due to gliding of the (3111) stacking fault in 6H SiC is predicted at 20 GPa under hydrostatic compression and at 18.8 GPa under shear deformation

The elastic constants for various SiC polytypes as a function of hydrostatic pressure were calculated following the procedure outlined in eqs. (4-6). Only pure SiC structures (no additives) were considered for determination of elastic constants. Similarly, only pure SiC structures (no additives)

were considered for calculation of shear strain applied along the $[01\overline{1}0]$ direction on (0001) plane in 2H, 3C, 4H, and 6H SiC. We observed that the stress-strain relations were nearly identical for all hexagonal polytypes up to $\gamma = 0.25$, and started to deviate above this value.

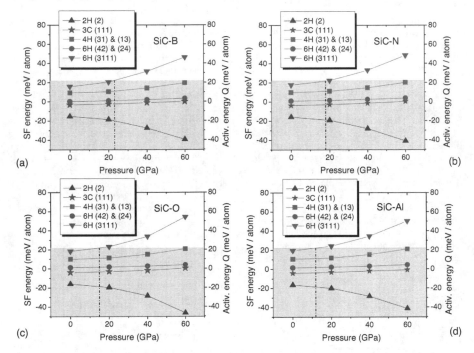

Figure 3. Stacking fault energy as a function of hydrostatic pressure for SiC polytypes with boron (a), nitrogen (b), oxygen (c) and aluminum (d) as substitutional impurities. The onset for plastic transformation due to gliding of the (3111) stacking fault in 6H SiC is predicted at 23 GPa for B, 18 GPa for N, 15 GPa for O, and 12 GPa for Al.

Knowledge of elastic constants allowed us to find lattice parameters of various SiC polytypes corresponding to each hydrostatic pressure value of 20 GPa, 40 GPa, and 60 GPa. Total energy calculations were performed at each lattice constants combination corresponding to a respective hydrostatic pressure value. Spin interaction energies were determined and finally the stacking fault energies were calculated following eq. (2). The results of this calculation are shown in Fig. 2a and Fig. 3. Addition of B leads to a decrease in the SFE gradient with respect to hydrostatic pressure. Inversely, addition of N, O, or Al is promoting an increase in the SFE pressure gradient for all SiC polytypes. As the result, the SFE with positive gradient (3C, 4H, and both 6H SF) are growing faster with hydrostatic pressure in the presence of N, O, or Al. In a similar way, the SFE with negative pressure gradients (2H) are falling faster with hydrostatic pressure in the presence of N, O or Al. The situation is reverse for B as substitute in the SiC lattice.

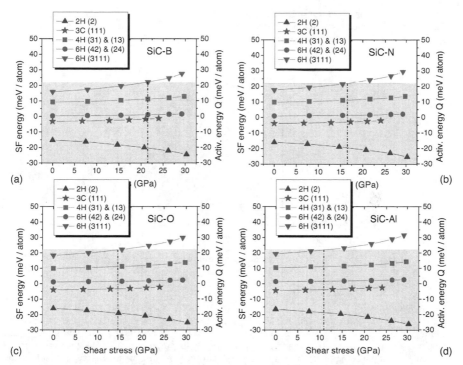

Figure 4. Stacking fault energy as a function of shear stress for SiC polytypes with boron (a), nitrogen (b), oxygen (c) and aluminum (d) as substitutional impurities. The onset for plastic transformation due to gliding of the (3111) stacking fault in 6H SiC is predicted at 21.7 GPa for B, 16.6 GPa for N, 14.6 GPa for O, and 11.0 GPa for Al.

In a similar fashion, total energy calculations and lattice optimizations were performed for each polytype for the values of shear stress corresponding to applied shear strains of $\gamma = 0.05, 0.10, 0.15, 0.20, 0.25, 0.30$. Spin interaction energies were determined and finally the stacking fault energies were calculated following eq. (2). The results of this calculation are presented in Figs. 2b and Fig. 4. Again, similarly to the hydrostatic pressure data, addition of B leads to a decrease in the SFE gradients with respect to shear stress, whereas the addition of N, O, or Al is increasing SFE shear stress gradients for all SiC polytypes.

The SFE as a function of hydrostatic pressure or shear stress was estimated against the activation energy Q for gliding in the SiC basal plane. As can be seen in Figs. 3-4, addition of N, O, or Al (in the concentrations considered here, specifically 1/24) decreases the onset for plastic deformation through gliding of the (3111) stacking faults in 6H SiC polytype, while the addition of B increases this critical pressure. The predicted SF-induced plasticity onsets in SiC under hydrostatic pressure and shear stress are shown in Fig. 5. As follows from analysis of Fig. 5, incorporation of oxygen, nitrogen, and, to the most extent, aluminum in the SiC structure significantly reduces SF-induced plasticity onset. In contrast, boron as a substitutional impurity increases SF-induced plasticity onset in SiC. Also, in all cases, SF-induced plasticity in SiC is predicted to start at lower stresses under shear deformation, as compared to purely hydrostatic compression.

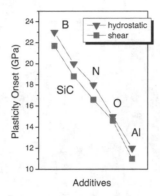

Figure 5. Calculated plasticity onset due to slipping along the 6H(3111) stacking fault in SiC with B, N, O, and Al as substitutional impurities under hydrostatic compression and shear deformation.

CONCLUSIONS

The (3111) stacking fault in 6H SiC is critical for SiC plasticity under external loading. Slipping along the stacking faults in SiC occurs at lower pressures under nonhydrostatic loading compared to purely hydrostatic compression. Doping with boron increases the pressures for SF-induced plasticity onset in SiC, whereas doping with nitrogen, oxygen and aluminum leads to decrease of the plasticity onset pressures. By expanding the pressure range for plastic deformation in silicon carbide under external loading to 11-12 GPa, aluminum as a substitutional impurity has the highest potential for the enhancement of ballistic performance of SiC based materials.

REFERENCES

1. D.P. Dandekar, A Survey of Compression Studies on Silicon Carbide (SiC), *U.S. Army Research Laboratory Report No. ARL-TR-2695* (2002).
2. G. Fanchini, private communication.
3. C. Cheng, R.J. Needs, and V. Heine, Inter-Layer Interactions and the Origin of SiC Polytypes, *J. Phys.* C, **21**, 1049-63 (1988).
4. P. Hohenberg, W. Kohn, Inhomogeneous Electron Gas, *Phys. Rev.*, **136**, B864-71 (1964).
5. W. Kohn, L. Sham, Self-Consistent Equations including Exchange and Correlation Effects, *Phys. Rev.*, **140**, A1133 -38 (1965).
6. M.H. Hong, A.V. Samant, P. Pirouz, Stacking Fault Energy of 6H-SiC and 4H-SiC Single Crystals, *Philos. Mag* A, **80**, 919-35 (2000).
7. L. Patrick, Inequivalent Sites and Multiple Donor and Acceptor Levels in SiC Polytypes, *Phys. Rev.*, **127**, 1878-80 (1962).
8. U. Lindefelt, H. Iwata, S. Öberg, P.R. Briddon, Stacking Faults in 3C-, 4H-, and 6H-SiC Polytypes Investigated by an Ab Initio Supercell Method, *Phys. Rev.* B **67** 155204 (2003).

9. J.C. Slater, *Quantum Theory of Molecular and Solids. Vol. 4: The Self-Consistent Field for Molecular and Solids* (McGraw-Hill, New York, 1974).

10. S.H. Vosco, L. Wilk, M. Nusair, Accurate Spin-Dependent Electron Liquid Correlation Energies for Local Spin Density Calculations: A Critical Analysis, *Can. J. Phys.*, **58**, 1200-11 (1980).

11. M.J. Frisch, G.W. Trucks, H.B. Schlegel, et al., Gaussian 03, Revision D.1 (Gaussian, Inc., Wallingford, CT, 2005).

12. W.J. Hehre, R.F. Stewart, J.A. Pople, Self-Consistent Molecular Orbital Methods I. Use of Gaussian Expansions of Slater Type Atomic Orbitals, *J. Chem. Phys.*, **51**, 2657-64 (1969).

13. J.B. Collins, P.v.R. Schleyer, J.S. Binkley, J.A. Pople, Self-Consistent Molecular Orbital Methods. XVII. Geometries and Binding Energies of Second-Row Molecules. A Comparison of Three Basis Sets, *J. Chem. Phys.*, **64**, 5142-51 (1974).

14. B. Mayer, H. Anton, E. Bott, et al., Ab-Initio Calculation of the Elastic Constants and Thermal Expansion Coefficients of Laves Phases, *Intermetallics*, **11**, 23-32 (2003).

15. A.T. Blumenau, C.J. Fall, R. Jones, et al., Structure and Motion of Basal Dislocations in Silicon Carbide, *Phys. Rev. B*, **68**, 174108 (2003).

16. M. Ikeda, H. Matsunami, and T. Tanaka, Site Effect on the Impurity Levels in 4*H*, 6*H*, and 15*R* SiC, *Phys. Rev. B*, **22**, 2842-54 (1980).

17. K. Karch, G. Wllenhofer, P. Pavone, et al. in *Proceedings of the 22nd International Conference on the Physics of Semiconductors, Vancouver, 1994*, edited by D. Lockwood (World Scientific, Singapore, 1995), p. 401.

18. P. Käckell, J. Furthmüller, F. Bechstedt, Stacking Faults in Group-IV Crystals: An Ab Initio study, *Phys. Rev. B*, **58**, 1326-30 (1998).

EVOLUTION OF THE ALN DISTRIBUTION DURING SINTERING OF ALUMINIUM NITRIDE DOPED SILICON CARBIDE

N. Ur-rehman*, P. Brown°,*, L.J. Vandeperre*
* UK Centre for Structural Ceramics and Department of Materials, Imperial College London, London SW7 2AZ, UK
° Defence Science and Technology Laboratory, Porton Down, Salisbury SP4 0JQ, UK

ABSTRACT

A study was carried out to determine how the distribution of aluminium nitride (AlN) changes during processing of silicon carbide (SiC) with AlN and carbon additions. Uni-axially pressed green bodies were formed from mixtures of AlN and SiC powders prepared by ball milling and rotary evaporator drying. The homogeneity of the distribution of the AlN powder in green bodies was studied by measuring the variability of the composition in small sized areas (16 μm^2) using the microanalysis facility attached to a scanning electron microscope. A secondary ion mass spectroscope attached to a focussed ion beam workstation (FIB-SIMS) was used to map their Al content.

Statistical analysis of this semi-quantitative data shows that the AlN powder is present in the powder mixture as dispersed, individual, particles after 24 hours of ball milling. It is shown that during vacuum hot pressing AlN redistributes towards SiC grain surfaces in the green body at temperatures as low as 1700 °C and that this coincides with the temperature at which sintering activity commences. When sintering is carried out rapidly, the high concentration of AlN on what appear to be SiC grain boundaries is preserved. For slower processing cycles, the AlN diffuses into the core of SiC grains and this is accompanied by formation of the 4H SiC polytype.

INTRODUCTION

The covalent nature of the bonding in silicon carbide (SiC) limits atomic mobility and necessitates the use of additives to promote densification. Additives that have been used include boron or aluminium with carbon[1, 2] and mixed oxides to promote the formation of a liquid phase[3, 4]. Aluminium nitride (AlN) has been used as a replacement for aluminium oxide in mixed oxide additive systems as it reduces weight losses during sintering. This enables SiC to be liquid phase sintered without the use of a powder bed[5]. Other authors have focussed on forming solid solutions of AlN and SiC[6-15]. In most of these cases yttria was added also to promote densification or alternatively sinterable SiC powders doped with boron were used.

In a recent extensive study by Ray and Cutler[16], SiC was hot pressed to high density using only AlN as an additive. While high densities where achieved using AlN additions of 2.5 wt.% and 5 wt.%, the start of densification varied considerably. The aim of this paper therefore is to investigate the role of AlN further by studying how its distribution changes during processing.

EXPERIMENTAL

Commercially available α-SiC and AlN powders, Table I, were mixed by ball milling using 10 mm diameter Si_3N_4 milling media (Union Process Inc., USA). Approximately 50 g of powder was mixed with 100 g methyl ethyl ketone and 75 g milling media. A phenolic resin (CR-96 Novolak, Crios Resinas, Brazil) was added as a source of carbon. The level of addition was determined to give 3 wt.% of carbon after pyrolysis of the resin. Following 24 h of ball milling, the powders were dried in a rotary evaporator (Rotavapor R-210/215, Büchi, Germany) and the phenolic resin converted to carbon through a 2 h heat treatment at 600 °C in flowing Ar gas.

The baseline sintering method consisted of hot pressing (FCT Systeme GmbH, Germany) 5 mm × 30 mm disks under vacuum at 2045 °C for 30 minutes with a maximum applied pressure of 21.2 MPa, Table II. To record changes in the AlN distribution during heating, for some samples the schedule was interrupted at 1700 °C and 1850 °C by switching off the power to the heating element once the temperature was reached and allowing the hot press to cool.

To separate out the effect of densification and time at temperature, a number of runs were carried out using a spark plasma sintering furnace (FCT Systeme GmbH, Germany) using 20 mm dies. In these experiments, the pressure (60 MPa), heating rate and sintering temperature (2100 °C) were increased, which allowed sintering the material to full density with only a 5 minute dwell at temperature. Full details of the schedule are given in Table III. Both instruments had a facility for measurement of the ram displacement during processing so that in-situ estimates of the relative density can be obtained.

Table I. Supplier data for the oxygen content, specific surface area, green density at a reference pressure and particle size of the powders used.

	α-SiC	AlN
Grade	H.C. Starck UF-15	H.C.Starck C
Oxygen content (wt.%)	1.5	0.1
Specific surface area (m^2 g^{-1})	14-16	4-8
Green density at 100 MPa (g cm^{-3})	1.55-1.75	
Particle size (μm)		
D 90%	1.20	2.30-4.50
D 50%	0.55	0.80-1.80
D 10%	0.20	0.20-0.35

Table II. Overview of vacuum hot pressing heating rates, dwell times and dwell temperatures.

Step	heating rate (°C min^{-1})	dwell (minutes)	Pressure (MPa)
<388 °C	n/a		8.5
388 °C - 1630 °C	50	0	8.5
1630 °C – 2030 °C	20	0	21.2
2030 °C – 2045 °C	5	30	21.2

Table III. Overview of spark plasma sintering heating rates, dwell times and dwell temperatures.

Step	heating rate (°C min^{-1})	dwell (minutes)	Pressure (MPa)
< 450 °C	(n/a)		16
450 °C - 1700 °C	100	0	16
1700 °C – 2100 °C	50	5	60

Sintered densities were determined using Archimedes' method. The polytypes of SiC were determined from x-ray diffraction data using the method of Ruska and coworkers[17]. The x-ray diffraction data was collected using a powder diffractometer using Cu Kα-radiation (PW 1729, Philips, The Netherlands) using a 2θ step size of 0.04° and 4 s per step.

In order to characterise the distribution of the AlN, a method inspired by the work of Zaspalis and Kolenbrander[18] was developed. This involved make at least 13 measurements of the composition of 16 μm² areas of each sample using the electron probe micro-analysis facility of a scanning electron microscope. If the typical size of AIN agglomerates is much smaller than the area analysed, then the standard deviation on the average of the measurements will be small as all measurements will typically sample the same number of agglomerates. However, as soon as the size of the analyzed region approaches the size of AIN agglomerates some measurements will include an AIN agglomerate whilst others may not and the standard deviation on the measurements will necessarily increase. Statistically, this approach is equivalent with sampling a binary population. The standard deviation therefore is given by:

$$\sigma = \sqrt{\frac{P(1-P)}{n}} \qquad (1)$$

where P is the fraction of one type of particles and n is the number of particles drawn from the population. Assuming a typical size (r) for the radius of AIN agglomerates and a planar powder packing fraction (f), the number of particles counted in the fixed area (A) depends on their radius:

$$n = \frac{f \cdot A}{\pi r^2} \qquad (2)$$

Therefore the standard deviation is predicted to vary with the size of the AIN agglomerates according to:

$$\sigma = \sqrt{\frac{\pi P(1-P)}{f}} \cdot \frac{r}{\sqrt{A}} \qquad (3)$$

and for a fixed area will be larger when the characteristic size of AIN particles is larger. As will be shown below this method works well as long as the AIN distribution is not too fine. Therefore, measurements of the distribution of Al concentration after sintering were also made using a secondary ion mass spectroscope attached to a focussed ion beam workstation (FIB-SIMS, FEI Company, USA). The sample was excited with a Ga beam of current 100 pA at zero tilt and the mass spectrometer counts for Al recorded while the beam was scanned over the surface.

RESULTS

The relative densities of vacuum hot pressed and spark plasma sintered (SPS) SiC powders containing 3.75 wt.% AIN and 3 wt.% C are plotted in Figure 1 as a function of temperature. In agreement with the observations made by Ray, Kaur and Cuttler[19], the higher heating rate associated with SPS shifts the densification to higher temperatures relative to the slower vacuum hot pressing cycle. However, it is clear that both processes yield similar, high, densities after sintering. Density measurements made after sintering are shown in Table IV. Similar traces were obtained for SiC powder containing 0.75 wt.% and 2.50 wt.% AIN. However a sample of the same SiC powder sintered without AIN only reached a relative density of 58% indicating that the AIN is crucial for densification to occur. The results of the measurements of the polytype content are shown in Table V.

Table IV also shows that the results for the standard deviation on the average composition measured over 16 μm² areas of the sample. For green bodies, the standard deviation is 1.2 wt%, which using eq.3 gives a characteristic particle size of 0.4 μm. This is of the same order of magnitude as the average particle size of the powder (0.8-1.8 μm) and therefore suggests that the mixing distributed the AIN in the body as individual particles. An Al map recorded for a green body SEM is shown in Figure

2. The size of the Al rich regions is again consistent with the AIN being present as individual powder particles.

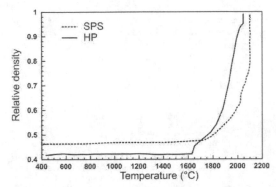

Figure 1. Variation of the relative density as estimated from the ram displacement of the hot press or SPS as a function of temperature for SiC containing 3.75 wt.% AIN and 3 wt.% C.

Table IV. Relative density as measured by Archimedes' method after sintering and the standard deviation on the average composition of 16 $\mu m2$ areas of the sample. The number in brackets indicates the number of measurements.

Sample	Relative density	Standard deviation on composition
Green pellet		1.2 wt% (13)
HP cycle interrupted at 1700 °C	0.58	0.7 wt% (20)
HP cycle Interrupted at 1850 °C	0.76	0.2 wt% (15)
Full SPS cycle	0.99	0.3 wt% (17)
Full HP cycle	0.97	

Table V. SiC Polytypes before and after sintering for various sintering schedules. All sintered samples contained 3.75 wt% AIN and 3 wt% carbon.

Sample	3C	4H	6H	15R
As received α-SiC powder	0%	6%	94%	0%
HP cycle interrupted at 1700 °C	0%	5%	95%	0%
HP cycle Interrupted at 1850 °C	0%	5%	95%	0%
Full SPS cycle	0%	7%	93%	0%
Full HP cycle	0%	22%	78%	0%

In the sample heated to 1700 °C only, the standard deviation is lower (0.7 wt.%). Hence, even at such a low temperature, the AIN distribution has become more homogeneous. Further heating reduces the standard deviation even further down to 0.2 wt.% - 0.3 wt.%. The latter is comparable to the standard deviation attributable to instrumental variation: measurements over much larger areas (1 mm^2) also gave standard deviations in the range 0.05 wt% - 0.2 wt%. Hence, the standard deviation due to instrumental limitations is larger than that due to Al homogeneity and hence for Al distributions at such fine scale this method can no longer be used.

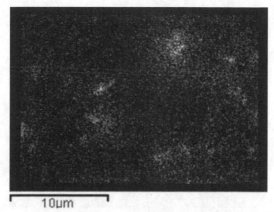

Figure 2. Al map for a pressed green pellet with 3.75 wt.% AlN: bright spots indicate the presence of Al.

The FIB-SIMS measurements of the Al distribution in sintered samples are shown in Figure 3. The Al maps of rapidly sintered samples with different levels of AlN, Figure 3a-c, all show a pattern where areas with zero counts (white) for Al are embedded in a matrix of material from which intermediate to high levels of Al counts are obtained. The shape and size of the regions without counts (white) suggest that these might be grains of SiC. It is also clear that the counts for Al increase as the AlN concentration in the sample increases. A further feature is that there are very fine particles giving much higher Al counts than the surrounding regions. These particles are much finer than the AlN particles originally present as can be ascertained by comparing with the regions giving high Al counts in the green pellet shown in Figure 2. Comparison of the maps shown in Figure 3c and Figure 3d shows that rapid processing leads to the Al remaining more concentrated in the regions between the grains, whereas in the case of the hot pressed sample the Al rich regions have become more diffuse.

DISCUSSION

Measurements of the distribution of Al in the samples reported here indicate that AlN starts to redistribute at ~1700 °C, Table IV, which coincides reasonably well with the temperature at which densification during hot pressing commences, Figure 1. Conversely, SiC samples without any AlN did not densify substantially indicating a strong connection between densification and AlN redistribution. The distribution of Al in sintered samples also indicates that this redistribution process is not by dissolution of the AlN particles in the surrounding SiC followed by bulk diffusion. This would lead to Al concentrations decreases monotonically from centres of high concentration. However the FIB-SIMS Al maps clearly show that it is distributed in veins surrounding regions with no counts for Al, Figure 3. This leaves diffusion along the surface of the SiC grains and evaporation-recondensation as mechanisms for redistribution of AlN. Even though it is difficult to determine with certainty which of these mechanisms contributes most to the redistribution, the small particle like pockets of high Al concentration are consistent with evaporation followed by recondensation in the necks and triple points between SiC grains. To confirm that this mechanism is possible the pressure-temperature phase diagram of pure AlN was calculated using the substance database of Thermo-Calc (Thermo-Calc Software, Sweden). As shown in Figure 4, under the vacuum conditions used, which is typically of the order of 1 mbar, AlN is predicted to decompose into aluminium and nitrogen gas from about 1700 °C onwards. Again, this is consistent with the temperature range where activation of the sintering occurs and where the SEM measurements indicate that the homogeneity of the Al is increasing.

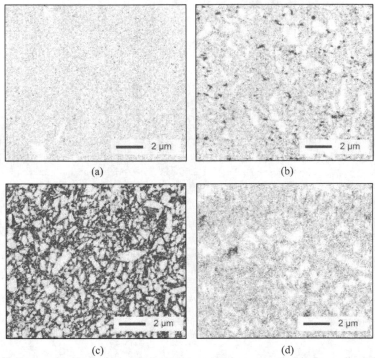

Figure 3. Focussed ion beam secondary ion mass spectroscopy maps showing normalised Al counts versus position (greyscales vary linearly between white : zero counts and black 10 or more counts for Al). The same grey scale was used for all maps to allow a more direct comparison between the images. (a) 0.75 wt% AlN, (b) 2.50 wt% AlN and (c) 3.75 wt% AlN densified using the rapid SPS cycle, and (d) 3.75 wt% AlN densified using the base-line hot-pressing cycle.

Surface or grain boundary diffusion then probably contributes further to the spreading of the AlN throughout the sample. The fact that the aluminium remains concentrated near the grain boundaries in rapidly densified samples indicates that actual alloying of SiC and AlN, through dissolution and diffusion of the AlN in the SiC is a much slower process which requires high temperatures and/or time. Indeed, in the slower hot pressing process, the decoration of the grain boundaries with Al becomes more diffuse, Figure 3d, and the XRD polytype measurements show that under these conditions the polytype content of the powder changes noticeably. Moreover, the observed change from 6H to 4H is consistent with AlN dissolution in the SiC described by Zangvil and Ruh[6] and other observations on the hot pressing of SiC with AlN additions[16, 19].

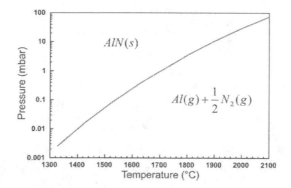

Figure 4. Pressure-Temperature phase diagram for pure AlN calculated using Thermocalc.

CONCLUSIONS

AlN additions as low as 0.75 wt.% allow SiC to be fully consolidated by pressure assisted densification. The onset of sintering in AlN doped SiC coincides with the temperature range where the distribution of Al in the sample was found to become more homogeneous whereas little densification occurred without the addition of AlN. This strongly indicates that the redistribution of the AlN activates the sintering. Detailed measurements of the Al distribution also suggest that it initially becomes concentrated in the triple points between SiC grains and on along grain boundaries. It is proposed that vapour transport of AlN is key to this redistribution process; an assertion supported by thermodynamic calculations which show, that for the processing conditions used to vacuum hot press SiC samples, evaporation of AlN is possible. Following this initial redistribution AlN then dissolves into the rim of SiC grains before moving towards their core by bulk diffusion. The polytype content of SiC samples only start to transform from 6H to 4H during this last step and requires that the material is left at temperature for extended times, i.e. rapid processing suppresses the transformation entirely. The polytype content of SiC samples can therefore be controlled via the use of AlN additions and the selection of an appropriate sintering schedule.

ACKNOWLEDGEMENTS

Naeem Ur-rehman and Luc Vandeperre thank the Defence Science and Technology Laboratory (Dstl) for providing the financial support for this work under contract No Dstlx-1000015398.

REFERENCES

1. S. Prochazka and R.J. Charles, Strength of Boron-doped hot-pressed SiC. *Ceramic Bulletin*, **52**, 885-891 (1973).
2. J.J. Cao, W.J. Moberleychan, L.C.D. Jonghe, C.J. Gilbert, and R.O. Ritchie, In situ toughened SiC with Al-B-C additions. *Journal of the American Ceramic Society*, **79**, 461-469 (1996).
3. M.A. Mulla and V.D. Krstic, Low Temperature Pressureless sintering of b-SiC with aluminium oxide and yttrium oxide additions. *Ceramic Bulletin*, **70**, 439-443 (1991).
4. R.A. Cutler and T.B. Jackson, Liquid phase sintered SiC. *Ceramic Materials and Components for Engines*, **42**, 309-318 (1994).
5. K. Strecker and M.-J. Hoffman, Effect of AlN-content on the microstructure and fracture toughness of hot-pressed and heat-treated LPS-SiC ceramics. *Journal of The European Ceramic Society*, **25**, 801-807 (2005).

6.   A. Zangvil and R. Ruh, Phase relationships in the silicon carbide - aluminium nitride system. *Journal of the American Ceramic Society*, **71**, 884-90 (1988).

7.   R. Ruh and A. Zangvil, Composition and Properties of Hot-Pressed SiC-AIN Solid Solutions. *Journal of the American Ceramic Society*, **65**, 260-265 (1982).

8.   S.-Y. Kuo and A.V. Virkar, Morphology of phase separation in AIN-Al2OC and SiC-AIN ceramics. *Journal of the American Ceramic Society*, **73**, 2640-2646 (1990).

9.   R. Huang, H. Gu, Z.M. Chen, T.H. Shouhong, and D.L. Jiang, The sintering mechanism and microstructure evolution in SiC-AIN ceramics studied by EFTEM. *International Journal of Materials Research*, **97**, 614-620 (2006).

10.   J. Lee, H. Tanaka, and H. Kim, Formation of solid solutions between SiC and AIN during liquid-phase sintering. *Materials letters*, **29**, 1-6 (1996).

11.   Y. Pan, In-situ characterization of SiC-AIN multiphase ceramics. *Journal of materials science*, **34**, 5357-5360 (1999).

12.   W. Rafaniello, K. Cho, and A.V. Virkar, Fabrication and characterization of SiC-AIN alloys. *Journal of Materials Science*, **16**, 3479-3488 (1981).

13.   J. Chen, Q. Tian, and A.V. Virkar, Phase separation in the SiC-AIN pseudobinary system: the role of coherency strain energy. *Journal of the American Ceramic Society*, **75**, 809-821 (1992).

14.   Q. Tian and A.V. Virkar, Interdiffusion in SiC-AIN and AIN-Al$_2$OC Systems. *Journal of the American Ceramic Society*, **79**, 2168-74 (1996).

15.   J.-F. Li and R. Watanabe, Preparation and mechanical properties of SiC-AIN ceramic alloy. *Journal of Materials Science*, **26**, 4813-4817 (1991).

16.   D.A. Ray, S. Kaur, R.A. Cutler, and D.K. Shetty, Effects of additives on the pressure-assisted densification and properties of silicon carbide. *Journal of the American Ceramic Society*, **91**, 2163-2169 (2008).

17.   J. Ruska, L.J. Gauckler, J. Lorenz, and H.U. Rexer, The quantitative calculation of SiC polytypes from measurement of X-ray diffraction peak intensities. *Journal of Materials Science*, **14**, 2013-2017 (1979).

18.   V.T. Zaspalis and M. Kolenbrander, Mixing homogeneity and its influence on the manufacturing process and properties of soft magnetic ceramics. *Journal of Materials Processing Technology*, **205**, 297-302 (2007).

19.   D.A. Ray, S. Kaur, and R.A. Cutler, Effect of additives on the Activation energy for sintering of silicon carbide. *Journal of the American Ceramic Society*, **91**, 1135-1140 (2008).

MICROSTRUCTURE, MECHANICAL PROPERTIES, AND PERFORMANCE OF MAGNESIUM ALUMINUM BORIDE (MgAlB$_{14}$)

Michael L. Whittaker and Raymond A. Cutler
Ceramatec, Inc.
2425 South 900 West
Salt Lake City, Utah, 84119

James Campbell and Jerry La Salvia
Army Research Laboratory
Aberdeen, Maryland 21005

ABSTRACT

Mg$_{0.78}$Al$_{0.75}$B$_{14}$, which is herein referred to as MgAlB$_{14}$, is a boride which has been studied in the laboratory but has not been tested ballistically previously due to the difficulty in making large components. This orthorhombic material consists of B$_{12}$ icosahedra and is reported to have high hardness like the rhombohedral B$_4$C in which similar covalent bonding occurs. The density of MgAlB$_{14}$ ($\approx$2.64 g/cc) is closer to B$_4$C (2.52 g/cc) than SiC (3.21 g/cc). While B$_4$C is preferred at lower threats, the ballistic performance of SiC is much better at higher threat levels. Hot pressed B$_4$C, MgAlB$_{14}$, and SiC were compared in the present work. The single-edged pre-cracked beam (SEPB) fracture toughness of MgAlB$_{14}$ was 3.4±0.4 MPa$\sqrt{m}$, which was intermediate between B$_4$C (2.2±0.5 MPa$\sqrt{m}$) and SiC-N (4.7±0.1 MPa$\sqrt{m}$). The fracture mode of magnesium aluminum boride was mostly transgranular, like B$_4$C, as opposed to the mainly intergranular fracture mode of SiC-N. The flexural strength of MgAlB$_{14}$ (390±37 MPa with a Weibull modulus of 11.7) was similar to B$_4$C (387±8.8 MPa with a Weibull modulus of 8.8), but much lower than that of SiC-N (558±50 MPa with a Weibull modulus of 14.5). The Vickers hardness values (at a one kilogram load) of all three materials (B$_4$C=26.0±2.0 GPa, SiC-N=22.5±0.8 GPa, and MgAlB$_{14}$=22.1±0.8 GPa) were much higher than that of the bullet (14.7 GPa) used for ballistic testing. The Young's modulus of MgAlB$_{14}$, which contained 4 wt. % MgAl$_2$O$_4$ as an impurity phase, was 397±1 GPa, which is lower than the other two materials (437±3 GPa for SiC-N and 436±2 GPa for B$_4$C). The V$_{50}$ ballistic performance of MgAlB$_{14}$ was approximately 250 m/s lower than SiC-N at the same areal density indicating that the material does not have promise for use at moderate or heavy threats.

INTRODUCTION

While there is much debate on what makes good armor it is universally agreed that low areal density (lightweight), high hardness (at least as hard as the projectile), and low cost (ceramic armor is expensive and is only used in limited applications) are important. The armor material of choice against steel-cored bullets is boron carbide (B$_4$C) due to its low areal density, while silicon carbide (SiC) is used against WC-cored bullets. While SiC (3.21 g/cc) has a higher density than B$_4$C (2.52 g/cc) it performs better at higher threats. Al$_2$O$_3$ (3.98 g/cc) is used due to its low cost and pressureless sintered materials are preferable to hot pressed materials when armoring tanks, due to the high volume of material that must be produced. Hardened steel is currently used to armor vehicles due to the cost of ceramic armor. In spite of the widespread use of steel, the ceramic armor market is substantial and fluctuates greatly based on need. It is difficult to find mechanical properties that correlate with ballistic performance, but ceramic

239

materials that perform well all have low porosity, high elastic modulus, and relatively high hardness.[1] Hardness and fracture toughness, however, do not correlate with the ballistic performance of SiC armor.[1-3]

A wide variety of borides exist with a spectra of interesting electrical, mechanical, thermal, and physical properties.[4] Matkovich and Economy identified $MgAlB_{14}$ as an orthorhombic structure (space group Imam) made up of B_{12} icosahedra with partial occupancy of Mg atoms, giving a theoretical density of 2.75 g/cc.[5] Further crystallography showed that about one quarter of the Al and Mg sites are vacant in the orthorhombic structure, leading to the formula $Mg_{0.78}Al_{0.75}B_{14}$, which results in a theoretical density of 2.59 g/cc.[6] The orthorhombic unit cell (a=5.844 Å, b=10.218 Å, and c=8.017 Å) has four molecules per unit cell as shown in Figure 1 resulting in a theoretical density of 2.64 g/cc.[7] The formula $MgAlB_{14}$ is used here for simplicity. Letsoala and Lowther have recently reviewed the structure of a variety of borides in an attempt to explain their properties.[7] They suggest that the average charge density between B atoms partially explains the high hardness of these materials. The B atoms lying outside the icosahedra donate part of their charge, which enhances the strength of the B-B bonds. Hardness for $MgAlB_{14}$ covers a range of values[7] partly due to the difficulty in measuring hardness and the different loads used. Single crystals have hardness in the range of 24-25 GPa.[8] High pressure densification of polycrystalline material resulted in 30-46 GPa hardness.[9] The hardness of aluminum magnesium boride is certainly in the range that would make acceptable armor.

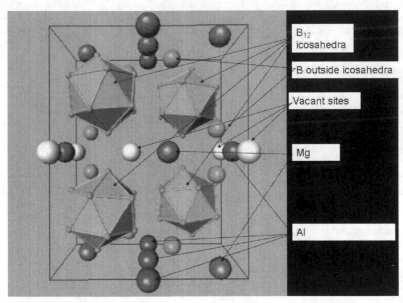

Figure 1. Structure of $Mg_{0.78}Al_{0.75}B_{14}$, where four B_{12} icosahedra occupy the orthorhombic unit cell at positions of (0,0,0), (0,0.5,0.5), (0.5,0,0) and (0.5,0.5,0.5), with the remaining eight B atoms located outside the icosahedra bonding. The Mg and Al atoms, which have four-fold coordination, are located at (0.25,0.359,0) and (0.25,0.75,0.25), respectively.[5-7,10]

Lee and Harmon[10] predicted high (470-509 GPa) Young's modulus, E, for MgAlB$_{14}$. Muthu, et al.[11] used high temperature X-ray diffraction to calculate a bulk modulus, K, between 196 and 264 GPa. The bulk and elastic moduli are related by $E=3K(1-2v)$, where v is Poisson's ratio. Taking a value of 0.1 for Poisson's ratio, gives a calculated Young's modulus in the range of 470 to 633 GPa. It is apparent that the material will have a high modulus and is of interest as a ballistic material.

There are no data for the ballistic performance of MgAlB$_{14}$, due to the fact that it has been difficult to produce. Single crystal growth or high-pressure densification of Mg, Al, and B has been the normal method for making the material. Bodkin[12] showed that dense MgAlB$_{14}$ could be produced by heating MgAlB$_{14}$ powder to 1600°C for one hour under 75 MPa pressure. The incorporation of Al and Mg in the unit cell lowers the densification temperature of MgAlB$_{14}$ by 600°C compared to B$_4$C, which is typically processed at 2200°C. New Tech Ceramics (Boone, IA), has recently been able to produce small tiles (50 mm x 450 mm x 5 mm) of MgAlB$_{14}$ by a proprietary process. Characterization of the material at Ceramatec resulted in properties as shown in Table 1. The fracture toughness and strength of MgAlB$_{14}$ are comparable to those of SiC-N but the elastic moduli is slightly lower than that of both SiC-N or solid state SiC when measured by the same technique.[1] The hardness is similar to that of SiC-N and lower than B$_4$C. Figure 1 shows a fracture surface of the material indicating that it fractures primarily transgranular, similar to B$_4$C and solid state SiC, but different than SiC-N. As Ceramatec is aware, there is no good method for predicting ballistic performance other than getting actual data.[1-3] SiC is the ceramic armor of choice for moderate to heavy threats due to the likely amorphitization of B$_4$C at high pressures. While mechanical properties look good for MgAlB$_{14}$, the question is whether it is at least comparable with hot pressed B$_4$C at the same areal density. Due to its lower processing temperature it has the potential to be a material of interest to the Army if the cost of processing the material were similar to boron carbide. This work was undertaken in order to ballistically test MgAlB$_{14}$ tiles in comparison with SiC and B$_4$C.

EXPERIMENTAL PROCEDURES

The three materials for this study were purchased by the Army Research Laboratory and provided to Ceramatec for characterization. The B$_4$C and SiC-N were hot pressed materials purchased from BAE Systems (Vista, CA) and are given the code of B and N, respectively. The MgAlB$_{14}$, given the code M in this paper, was purchased from New Tech Ceramics (Boone, IA) and no processing details are available. The thickness of the materials supplied were 11.4 mm (material B), 15.5 mm (material M), or 25.4 mm (material N). The billets were sliced and then ground with a 180 grit diamond wheel to make 3 mm x 4 mm x 45 mm bars as specified by ASTM C-1421-99. Density was measured by water displacement. Fracture toughness was measured using the single-edge precracked beam (SEPB) technique[13] as described previously[14]

Table I
Properties of New Tech MgAlB$_{14}$ Measured at Ceramatec

Density (g/cc)	SEPB Toughness (MPa-m$^{1/2}$)	Elastic Modulus (GPa)	Flexural Strength (MPa)	HV1 (GPa)
2.64	4.2±0.3	396±3	516±76	23.6±0.5

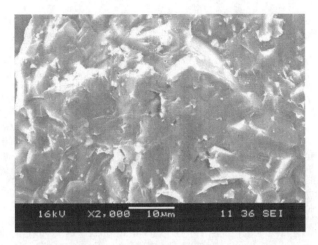

Figure 2. Fracture surface of New Tech's MgAlB$_{14}$ showing mainly transgranular fracture. See properties of this material in Table I.

except that black printer ink was used to mark the crack location. All crack planes were parallel to the hot pressing direction. Each data point is the mean of 5 bars tested, with error bars representing two standard deviations.

A microhardness machine (Leco model LM-100) was used to obtain Vickers and Knoop hardness data on polished SEPB bars. Data were taken at a load of 9.8 N. Each data point represents the mean of ten measurements, with error bars representing two standard deviations. Rietveld analysis[15,16] was used to quantify phases or polytypes present in the materials with X-ray diffraction patterns collected from 20-80° 2Θ, with a step size of 0.02°/step and a counting time of 4 sec/step.

Polished samples were etched to reveal their grain boundaries. Material M was etched in a modified Murakami solution[17] at 80°C for 60 seconds. Materials B and N were thermally etched at 1550°C in flowing Ar for one hour. Grain size was determined by the line-intercept method, where the multiplication constant was 1.5 (equiaxed grains).[18] Approximately 500 grains were measured for each composition in order to get a mean grain size. The aspect ratios of the three most acicular grains in each of 5 micrographs were used to estimate a comparative aspect ratio.

The fracture mode was determined from polished, precracked SEPB bars. The precracked bars were subsequently etched as described above to get a quantitative estimate of the fracture mode by viewing the crack path over a distance of 150-650 μm, depending on grain size.

Flexural strength was measured on 25 bars (3 mm x 4 mm x 45 mm) using a 40 mm support span and a 20 mm loading span, with the crosshead speed at 0.5 mm/min. A two-parameter Weibull analysis was used to calculate the characteristic strength. Young's modulus was measured in flexure using strain gages.

Ballistic testing was performed at ARL on 100 mm x 100 mm tiles using steel to surround the targets and composite backing and cover plates. The M (thickness of 15.5 mm) and

N (thickness of 12.8 mm) materials were tested using the same technique at the same areal density. The V_{50}, in theory, is the velocity of the bullet at which the probability of the projectile penetrating through the composite backing plate is 50%. Due to the limited number of targets tested, this value was taken as the mean of the two highest velocity tests at which the bullet did not fully penetrate the composite backing and the two lowest velocity tests at which the bullet fully penetrated the backing.

RESULTS AND DISCUSSION

Materials Characterization

Table II gives density, SiC polytypes, mean grain size, aspect ratio, Young's modulus, as well as other phases identified by XRD for the three materials described in Table I. The N material was similar to what has been reported previously for this material, consisting primarily of the 6H polytype, with minimal porosity and high Young's modulus.[1] The B material had lower density than would be expected for a hot pressed material and consisted of a variety of boron carbide phases, as evidenced by the asymmetric peaks (see Figure 3(a)). While free carbon is used as a sintering aid, resulting in the graphite found in the microstructure, the aluminum oxynitride and hexagonal boron nitride phases were unexpected. No Rietveld fitting was attempted for Material B. The low Young's modulus measured is indicative of the porosity in the material and the additional phases present. Material M consisted of 95.6 wt. % $Mg_{0.78}Al_{0.75}B_{14}$, 3.9 wt. % $MgAl_2O_4$, and 0.5 wt. % Al (see Figure 3(b)). No FeB or W_2B_5 were present, as had been reported by other researchers.[9,12] The lattice parameters for the $Mg_{0.78}Al_{0.75}B_{14}$ phase were a=5.8491±0.0003 Å, b=10.3171±0.0006 Å, and c=8.1175±0.0004 Å resulting in a theoretical density of 2.58 g/cc for the $Mg_{0.78}Al_{0.75}B_{14}$ phase, similar to the value reported by Higashi and Ito.[6] Using the Rietveld fit, the theoretical density of the M material was calculated to be 2.61 g/cc, which is lower than the measured value. The theoretical density of material M is therefore unknown, but there is little porosity in the material (see Figure 4). The modulus is similar to that measured at Ceramatec previously (see Table 1) and is lower than what was predicted for this material. The presence of the spinel and aluminum phases lowers the modulus, which is similar to some solid-state sintered silicon carbides. These pressureless sintered SiC materials perform reasonably well ballistically against moderate threats. The modulus of the B material was similar to N, likely due to the porosity and secondary phases in the B material.

The N material is the finest-grained of the three materials, all of which are primarily equiaxed (see Figure 5). The N material fractures mostly intergranularly, which is apparent on both fracture surfaces (see Figure 6) and with hardness indentations on polished surfaces, as shown in Figure 7. The M material has a grain size which is smaller than the boron carbide. The $MgAl_2O_4$ phase, which likely forms due to the oxygen adsorbed on the starting materials,[12] is

Table II

Characterization of Materials

Designation	Density (g/cc)	SiC Polytypes 4H	6H	15R	Other Phases or Polytypes	Grain Size (μm)	Aspect Ratio	E (GPa)
B	2.47±0.02	Not aplicable			Al_3O_3N, BN, C	9.8±0.7	2.8±0.1	436±2
M	2.62±0.01	Not aplicable			$MgAl_2O_4$, Al	4.3±0.4	2.4±0.2	397±1
N	3.22±0.01	2.1	92.3	4.1	2H=0.1, 3C=1.4	3.2±0.2	2.7±0.4	437±3

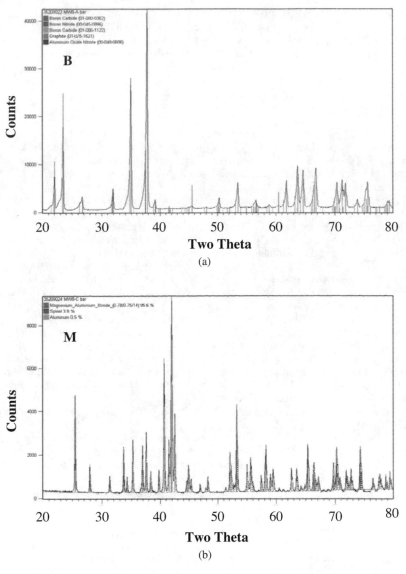

Figure 3. X-ray diffraction patterns for materials (a) B and (b) M. Rietveld fit shown for M.

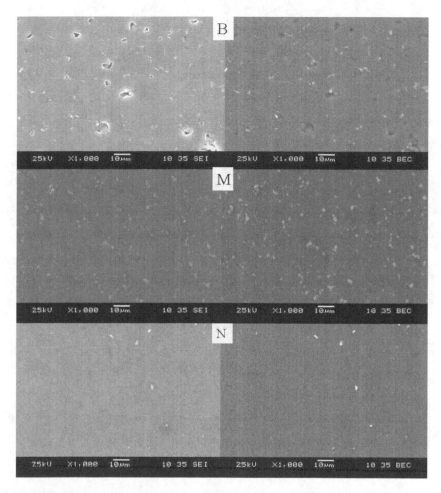

Figure 4. SEM images (secondary on left and backscattered on right) of polished surfaces. Markers are 10 µm.

located at triple points and grain boundaries, although transmission electron microscopy would be required in order to see if a continuous grain boundary phase exists. Material M appears microstructurally to be a ceramic that could perform well ballistically, as it has little porosity and only a small amount of secondary phase present in the material. The apparent porosity on the polished surface is not indicative of the porosity in the material. The B material looked the most porous of the three hot pressed materials, with porosity apparent on fracture surfaces (see Figure 5) and pullout on polished surfaces (see Figure 6).

Table III
Mechanical Property Comparison

| Designation | Strength (MPa) | | | % Intergranular | Toughness | Hardness (GPa) | |
	Mean	Char.[a]	m[b]	Fracture	(MPa-m$^{1/2}$)	HK1	HV1
B	387±64	411	8.8	8	2.2±0.5	19.7±1.0	26.0±2.0
M	390±38	407	11.7	6	3.4±0.4	18.8±0.4	22.1±0.8
N	558±50	579	14.5	72	4.7±0.1	19.5±0.5	22.5±0.8

a. Characteristic strength (63.2 % probability of failure).
b. Weibull modulus.

Table III gives mechanical properties for bars cut out of the 100 mm x 100 mm billets. The N material is similar to what has been reported previously.[1] The fracture toughness and strength of the M material were not has high as had been expected, based on the properties evaluated previously (see Table 1). This is the first time to the authors' knowledge that large MgAlB$_{14}$ plates have been prepared. The Weibull modulus (see Figure 8) was very acceptable for this material with strength similar to pressureless sintered SiC. The fracture toughness of material M is considerably higher than that of pressureless sintered silicon carbide, which is 2.5 MPa√m when measured by this same technique. The M material was not as hard as the B material, but comparable in Vickers hardness to material N. The fracture toughness values of the materials are not highly correlated with the amount of intergranular fracture, as had been expected.

The reason for the difference in mechanical properties of MgAlB$_{14}$ for the small plates tested previously (see Table I) and the larger plates tested in this work (see Table III) is not related to density or phases present, as XRD patterns were similar for both materials. Further characterization of these materials would be necessary to explain their difference in toughness and strength.

Ballistic Testing

Only materials M and N were ballistically tested. The initial testing of material M was at the same velocity as the V$_{50}$ of material N. The test velocity was successively dropped until partial values were obtained. The V$_{50}$ of material M is not well quantified, but is approximately 250 m/s below that of material N. No characterization of ballistic debris or TEM work was

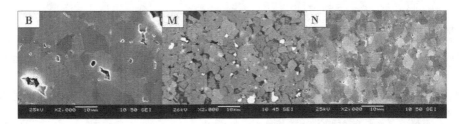

Figure 5. Polished and etched cross-sections. Markers are 10 μm.

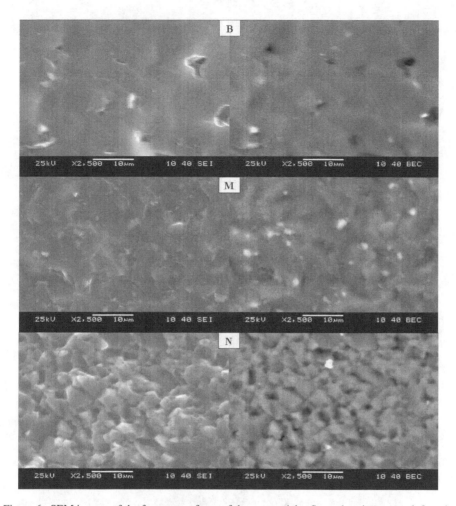

Figure 6. SEM images of the fracture surfaces of three materials. Secondary images on left and backscattered imaging on right. Markers are 10 μm. Note that B and M fracture primarily transgranularly while N fractures intergranularly. Light phase in backscattered imaging of M is the spinel phase.

performed so it is difficult to speculate on the reasons for the poor performance of this material. It is not entirely unexpected, however, as B_4C does not perform well when tested at moderate to heavy threats. If phase pure material M can be produced, it should be tested, since secondary

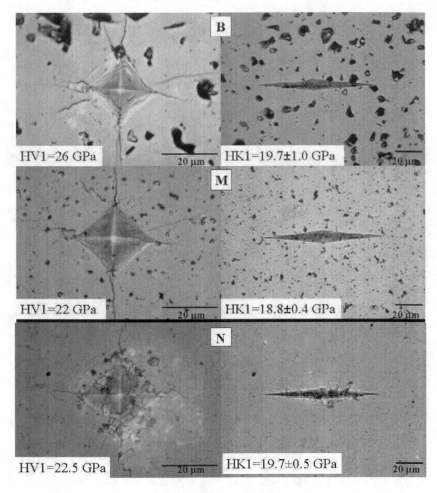

Figure 7. Hardness indents at one-kilogram loads. Markers are 20 μm. Intergranular fracture of N material is readily apparent on polished and indented surfaces.

phases can influence ballistic performance. It is apparent that $MgAlB_{14}$ performs not only much worse than SiC-N, but also is much worse than pressureless sintered SiC, which is within 10 % of the V_{50} value of SiC-N. Since cost is a big driver for ballistic materials, further efforts should not be directed at using this material in armor applications since B_4C is already a commodity material and $MgAlB_{14}$ armor can not be produced at similar cost.

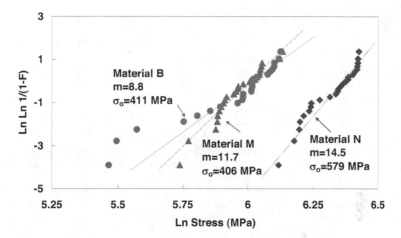

Figure 8. Weibull plots for the three materials tested.

CONCLUSIONS

The V_{50} performance of $MgAlB_{14}$ was approximately 250 m/s lower than that of SiC-N, a standard material used in ballistic tests, at the threat level investigated when identical areal densities were compared. The poor ballistic performance of $MgAlB_{14}$, coupled with its high cost, militates against its use as an armor material. While the mechanical properties of $MgAlB_{14}$ are similar to, or exceed those of pressureless sintered SiC, it performs much worse ballisitically at a moderate threat level. This demonstrates again the importance of performing ballistic tests in order to evaluate a material. The results of this work are not entirely surprising in light of the known poor performance of B_4C, which is also made by bonding B_{12} icosahedra.

ACKNOWLEDGEMENT

Appreciation is expressed to Sarbjit Kaur, Ben Isaac, and Lyle Miller for their help.

REFERENCES
[1] D. Ray, R. M. Flinders, A. Anderson, R. A. Cutler, J. Campbell, and J. W. Adams, "Effect of Microstructure and Mechanical Properties on the Ballistic Performance of SiC-Based Ceramics," *Ceram. Eng. Sci. Proc.*, **27**[7] 85-96, (2006).

[2] D. Ray, R. M. Flinders, A. Anderson, R. A. Cutler and W. Rafaniello, "Effect of Room-Temperature Hardness and Toughness on the Ballistic Performance of SiC-Based Ceramics," *Ceram. Eng. Sci. Proc.*, **26**[7], (2005).

[3] R. Marc Flinders, D. Ray, A. Anderson and R. A. Cutler, "High-Toughness Silicon Carbide as Armor," *J. Am. Ceram. Soc.* **88**[8], 2217-26 (2005).

[4]R. A. Cutler, "Engineering Properties of Borides," pp. 787-803 in Engineered Materials Handbook, Vol. 4: Ceramics and Glasses (ASM, Materials Park, PA 1991).

[5]V. I. Matkovich and J. Economy, "Structure of $MgAlB_{14}$ and a Brief Critique of Structural Relationships of Higher Borides," Acta. Cryst., **B26** 616-21 (1970).

[6]I. Higashi and T. Ito, "Refinement of the Structure of $MgAlB_{14}$," J. Less Common Metals, **92**[2] 239-46 (1983).

[7]T. Letsoala and J. E. Lowther, "Systematic Trends in Boron Icosahedral Structured Materials," Physica B, **403** 2760-67 (2008).

[8]S. Okada, K. Kudou, T. Mori, T. Shishido, I. Higashi, N. Kamegashira, K. Nakajima, and T. Lundström, "Crystal Growth of Aluminum Magnesium Borides from Al-Mg-B Ternary System Solutions and Properties of the Crystals," Mater. Sci. Forum, **449-452**, 315-368 (2004).

[9]B. A. Cook, J. C. Harringa, and A. M. Russell, "Processing Studies and Selected Properties of Ultra-Hard $AlMgB_{14}$," J. Adv. Mater., **36**[3] 56-63 (2004).

[10]Y. Lee and B. N. Harmon, "First Principles Calculation of Elastic Properties of $AlMgB_{14}$," J. Alloys Compounds, **338**[1-2] 242-247 (2002).

[11]D. V. S. Muthu, B. Chen, B. A. Cook, and M. B. Kruger, "Effects of Sample Preparation on the Mechanical Properties of $AlMgB_{14}$," High Pressure Res., 28[1] 63-68 (2008).

[12]R. Bodkin, "A Synthesis and Study of $AlMgB_{14}$," Ph.D. Dissertation, University of the Witwatersrand (Johannesburg, South Africa, 2006).

[13]ASTM C 1421-99, Standard Test Methods for Determination of Fracture Toughness of Advanced Ceramics at Ambient Temperature, pp. 641-672 in 1999 Annual Book of Standards (ASTM, Philadelphia, PA 1999).

[14]D. Ray, M. Flinders, A. Anderson, and R. A. Cutler, "Hardness/Toughness Relationship for SiC Armor," Ceram. Sci. and Eng. Proc., **24**, 401-10 (2003).

[15]H. M. Rietveld, "A Profile Refinement Method in Neutron and Magnetic Structures," J. Appl. Crystallogr., **2**, 65-71 (1969).

[16]D. L. Bish and S. A. Howard, "Quantitative Phase Analysis Using the Rietveld Method," J. Appl. Crystallogr., **21**, 86-91 (1988).

[17]D. H. Stutz, S. Prochazka and J. Lorenz, "Sintering and Microstructure Formation of b-Silicon Carbide," J. Am. Ceram. Soc., **68**[9], 479-82 (1985).

[18]E. E. Underwood, Quantitative Stereology, (Addison-Wesley, Reading, MA. 1970).

MICROSTRUCTURAL DEVELOPMENT AND PHASE CHANGES IN REACTION BONDED BORON CARBIDE

P. G. Karandikar, S. Wong, G. Evans, and M. K. Aghajanian
M Cubed Technologies, Inc.
1 Tralee Industrial Park
Newark, DE 19711

ABSTRACT
    Reaction bonded silicon carbide (RBSC) and reaction bonded boron carbide (RBBC) materials have been used successfully for armor applications over the last decade. Silicon carbide is inert and does not react with molten silicon during the reaction bonding process. Boron carbide however, can react with molten silicon during the reaction bonding process. While this reaction can lead to undesirable reaction zones, cracking and yield losses during manufacturing, the same reaction may be harnessed to create lightweight phases in the microstructure that offer better mechanical, thermal and ballistic properties in-lieu of the residual silicon that has much poorer properties. Addition of alloying elements allows further tailoring of the reaction-synthesized phases. This study focuses on development of such phases in RBBC with the eventual goal of improving mechanical, thermal and ballistic properties. Detailed studies are undertaken to drive the reactions to various levels and characterize the microstructure and phases (optical, SEM, and TEM) in the resultant composites. It is found that the composition and phase structure of the starting B_4C changes with processing time and alloying.

INTRODUCTION
    Sintered and hot pressed Al_2O_3, B_4C and SiC have been traditionally used in personnel armor systems. Reaction bonded SiC (RBSC) and reaction bonded B_4C (RBBC) have been reported in the literature as far back as 1940s[1-3]. However, their use in armor systems started in the late 1990s[4-7]. B_4C is the lowest density and highest hardness ceramic typically used for personnel armor. While RBBC offers good performance for currently fielded body armor, there is a continuous desire to obtain the same performance with lighter weight systems.
    In the reaction bonding process (silicon-based matrices), good wetting and highly exothermic reaction between liquid silicon and carbon is utilized to achieve pressure-less infiltration of a powder preform. This process is given many names such as reaction-bonding, reaction-sintering, self-bonding, and melt infiltration. A schematic of the reaction bonding process is shown in Figure 1. The steps in the process are as follows (1) Mixing of B_4C (or SiC) powder and a binder to make a slurry, (2) Shaping the slurry by various techniques such as casting, injection molding, pressing etc., (3) Drying and carbonizing of binder, (4) Green machining, (5) Infiltration (reaction bonding) with molten Si (or alloy) above 1410°C in an inert/vacuum atmosphere, and (6) Solidification and cooling. During the infiltration step, carbon in the preform reacts with molten Si forming SiC around the original ceramic particles, bonding them together – hence the term reaction bonding. During the development and manufacture of RBBC based composites for armor and other applications, it has been found that a variety of microstructural features occur in RBBC due to the reactivity of B_4C with molten Si and its alloys. Examples of these microstructural features are shown in Figure 2. It is postulated that some the phases formed during RBBC processing, including those formed when the Si is alloyed

251

with elements such as Al, Ni, could lead to better, lighter ballistic materials. This study focuses on systematic creation and analysis of these phases.

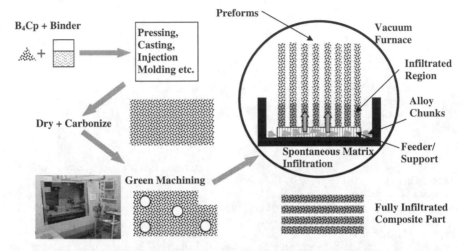

Figure 1. Schematic representation of the reaction bonding process.

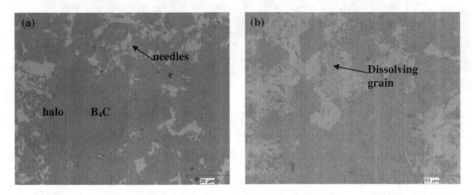

Figure 2. Microstructural features found in RBBC materials (a) Halo (lighter periphery) found around B$_4$C grains, and needle shaped reaction products (b) B$_4$C grain in the process of dissolution

EXPERIMENTAL PROCEDURE

RBBC plates (100 x 100 x 10 mm) were made with four different infiltration times (60, 120, 240 and 360 minutes). For all plates, preforms were made by the process steps outlined in Figure 1. RBBC plates were also made with alloying of the infiltrant Si with Al and Ni. Microstructure and phases in these materials were characterized by optical microscopy, SEM,

EDAX, and X-ray diffraction (XRD). The starting boron carbide powder was also analyzed by XRD to establish the starting phase composition. XRD was used to determine the changes in the phase of the starting boron carbide due to varying processing conditions. Each comminuted sample (WC ball mill) was placed into a standard sample cup and put into a Panalytical X'Pert MPD Pro diffractometer using Cu radiation at 45KV/40mA over the range of 20° - 80° with a step size of 0.0156° and a count time of 500 seconds per step. The phases were then identified by use of the Powder Diffraction File published by the International Centre for Diffraction Data. The detected phases were quantified with the aid of a Rietveld analysis. The measured lattice parameters were used to determine the B to C ratio in the boron carbide. One sample showing needle-type reaction phase was subjected to TEM. Densities (Archimedes principle, ASTM B311), elastic moduli (ultrasonic pulse-echo, ASTM E494-05) of the resultant materials were also characterized.

RESULTS AND DISCUSSION

The processing conditions for various samples and their properties are summarized in Table I. As can be seen from the data, increasing the processing time did not systematically or significantly affect density and elastic modulus. Samples with Al and Ni alloying show slightly higher densities due to the higher densities of Al and Ni compared to Si. Optical micrographs of various samples produced in this study are shown in Figures 3. All samples show B_4C grains (nominally) and Si (or alloy matrix region). The reaction formed SiC is not discernible in low magnification optical micrographs. In the samples containing Al or Ni alloying, smaller boron carbide grains appear to fuse together making large islands of ceramics. Also, brighter Al or Ni phases can be seen within the matrix phase (Si).

Table I. Processing conditions and properties of sample plates

No.	ID	Matrix	Temperature (°C)	Time (min)	Density (g/cc)	Young's Modulus (GPa)
1	030409VU-2	Si	1530	60	2.55	384
2	022209VT-3	Si	1530	120	2.55	389
3	030109VU-11	Si	1530	240	2.56	379
4	030309VU-2	Si	1530	360	2.56	375
5	031209VU-5	Si - 10 Al	1530	300	2.60	402
6	121109VP2-5	Si - 8 Ni	1530	300	2.62	374

The XRD patterns of various samples are shown in Figure 4 along with the XRD pattern of the starting B_4C powder. The phase composition determined by the Rietveld analysis is shown in Table II. The phase compositional data obtained by XRD analysis suggests the following:
(1) All samples show reaction formed SiC – as is expected in the reaction bonding process.
(2) At low process times, only β-SiC is formed. At longer process durations however, some α-SiC is also formed.
(3) The content of α-SiC increased as the process time increased.
(4) It is important to note here that while SiC is a line compound (1:1 Si to carbon atomic ratio), B_4C is not a line compound and can exist over a range of compositions, from 9 atomic percent carbon to 20 atomic percent carbon (B_4C) as a solid solution[8]. For the nominally

B_4C phase identified in this work, the lattice parameters were determined to be $a_o = 5.6061 \pm 0.0003$ Å and $c_o = 12.0867 \pm 0.0007$ Å. Next, using these lattice parameters and the correlations reported in reference 9, the carbon content was determined to be 18.4% (atomic). Hence, the composition of this phase was determined to be $B_{4.44}C$.

(5) At short process durations, part of the $B_{4.44}C$ is converted to $B_{4.44-x}Si_xC$. This is clearly evident from the gradual elimination of the $B_{4.44}C$ peak at ~37.6 degrees (2θ) and appearance of new $B_{4.44-x}Si_xC$ characteristic peak at ~37.5 degrees. At longer process durations, all the $B_{4.44}C$ is converted to $B_{4.44-x}Si_xC$.

(6) In the case of Al alloying, all the $B_{4.44}C$ is converted to an aluminum-boro-carbide ($AlB_{24}C_4$ or $B_{12}Al_{0.5}C_2$) for the 300 minutes of processing time.

TEM, SEM, EDAX and ESCA Analysis

One of the samples that showed presence of significant "needle formation" was subjected to transmission electron microscopy. A small sample was cut using a field ion beam (FIB). This sample was then thinned down using ion milling to fabricate a thin foil for TEM. Figure 5a shows TEM micrographs obtained form this foil. Here, a central grain surrounded by features of various shapes (twinned grain, needles) can be seen. Higher magnification images of some of the features are shown in Figures 5c and 5d. The same foil was also observed in SEM under the back scattered electron mode (atomic weight contrast) and the corresponding image is shown in Figure 5b. EDAX analysis was conducted on the features in image 5b. The detectability of light elements such as B and C has some limitations in EDAX. Also, typical analysis volume is few micrometer cubed. Thus, there are contributions from the surroundings for features that are very small. In spite of these limitations, interesting results were obtained. At location (1), elemental composition was predominantly boron and carbon with very minor amount of Si. Thus, in agreement with the XRD analysis, this is likely to be the $B_{4.44-x}Si_xC$ phase. At location (2) and (3), the predominant composition was Si and carbon suggesting that these needle-shaped features are reaction formed SiC. ESCA analysis was also performed on the sample showing needles and confirmed the composition of needles to be SiC.

The phase composition data presented in Table II were analyzed in terms of fraction of $B_{4.44-x}Si_xC$ formed as a function of process time and are plotted in Figure 6a. This plot suggests that the conversion of a $B_{4.44}C$ grain to $B_{4.44-x}Si_xC$ is a time dependent phenomenon. Thus, the core-rim structure observed in the $B_{4.44}C$ grains can be explained based on this time dependent conversion. Figure 6b shows a schematic of this structure and the process. The shift of the key peaks in XRD to lower angular values (Figure 4) suggests "swelling" of the lattice. Interestingly, during processing of RBBC components a small expansion is found to occur going from the preform stage to the infiltrated stage. This expansion also increases with increasing process time or temperature. When Al is added to the Si infiltrant, Al insertion into the $B_{4.44}C$ lattice occurs preferentially over Si insertion.

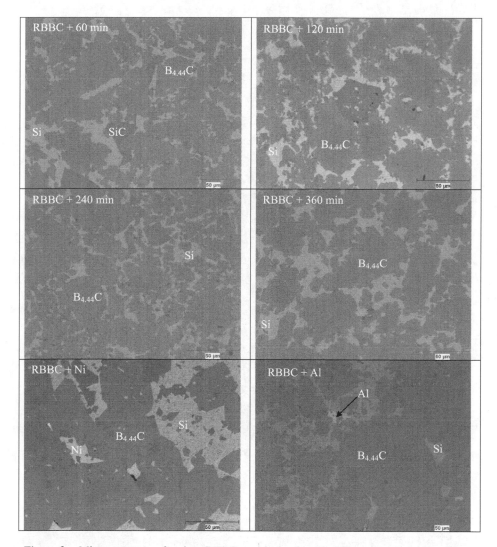

Figure 3. Microstructures of various RBBC samples made with different processing conditions and alloying.

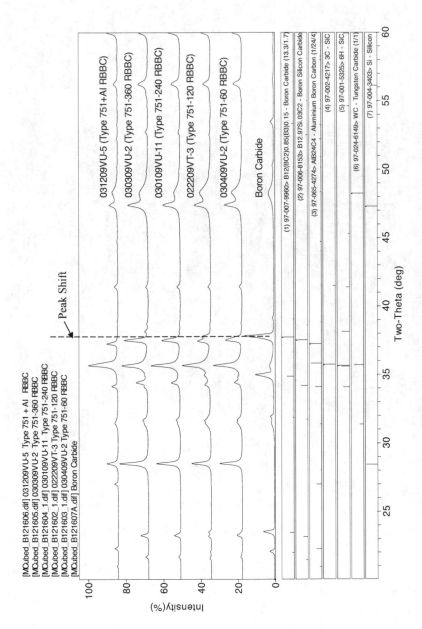

Figure 4. Comparison of XRD patterns of the starting B₄C powder and various RBBC samples.

Author Index